E. M. Reynolds

Modern Methods In Elementary Geometry

E. M. Reynolds

Modern Methods In Elementary Geometry

ISBN/EAN: 9783741150999

Manufactured in Europe, USA, Canada, Australia, Japa

Cover: Foto ©Thomas Meinert / pixelio.de

Manufactured and distributed by brebook publishing software
(www.brebook.com)

E. M. Reynolds

Modern Methods In Elementary Geometry

MODERN METHODS

IN

ELEMENTARY GEOMETRY.

BY

E. M. REYNOLDS, M.A.

MATHEMATICAL MASTER IN CLIFTON COLLEGE, MODERN SIDE.

London and Cambridge:
MACMILLAN AND CO.
1868.

PREFACE.

GEOMETRY has received extensive developments in modern times, but in England there has been no corresponding improvement in elementary teaching. Our principal and only text-book is a work written more than two thousand years ago.

On the Continent, however, great attention has been given to the form in which the science should be taught. New principles have been introduced, old ones have received an extended application, useless restrictions have been abandoned, and the style of disputation has given place to that of enquiry.

Hence that superiority of results which has lately come so prominently before the public.

Some change, it is evident, in our English ways of teaching can now no longer be postponed, and this little book, mainly derived from French and German sources, has been written in the hope of facilitating that change. It has been

constructed on one plan throughout, that of always giving in
the simplest possible form the direct proof from the nature of
the case. The axioms necessary to this simplicity have been
assumed without hesitation, and no scruple has been felt as to
the increase of their number, or the acceptance of as many
elementary notions as common experience places past all
doubt. For in an experimental science we are bound to have
recourse to nature for as many principles as are necessary to
the clearest exposition of that science, subject only to the
conditions that these principles be well chosen and beyond
reasonable dispute.

The book differs most from established teaching in its
constructions, and in its early application of Arithmetic to
Geometry.

The arbitrary restrictions of Euclid involve him in various inconsistencies, and exclude his constructions from use.
When, for instance, in order to mark off a length upon a
straight line, he requires us to describe five circles, an equilateral triangle, one straight line of limited, and two of unlimited length, he condemns his system to a divorce from
practice at once and from sound reason. The constructions
given in this book are theoretically consistent, and are employed by practical men.

In the application of arithmetical methods, all that has
been done has been to anticipate a step which sooner or

later is necessary to progress. Valuable results are seldom attained without the bold and skilful combination of the best means at our command, and it is to such a combination that the range and power of Modern Geometry are mainly due. Measurement is besides indispensable to every practical enquiry, and it seems unwise to deprive the learner of the power and interest given by its employment.

Moreover some method of dealing with questions of proportion is absolutely necessary to replace the abandoned Fifth book of Euclid, and to prevent the confusion inseparable from the present thoroughly irrational manner of studying the Sixth book.

Continental practice sanctions the innovations which the Author has ventured to adopt.

As to the book itself, though so small it contains all that is usually read of the first six books of Euclid, with considerable additions. It has been designed as a sufficient introduction to Trigonometry, Conic Sections, and applied Mathematics. It is not intended for private reading, but for a class-book, and the master in using it should work into his teaching the theorems and constructions given at the end of each part. These have been selected with care, and are not mere enigmas but propositions of some importance either in principle or application. The time spent upon them will certainly not be wasted.

MODERN METHODS

IN

ELEMENTARY GEOMETRY.

BOOK I.

THE STRAIGHT LINE.

Two straight lines AB, AC which meet form an angle.

The straight lines AB, AC are called the sides of the angle, and the point A the vertex.

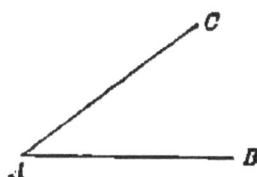

An angle is usually distinguished by three letters, as BAC, CAD, DAB, of which the one at the vertex must be placed between the other two. When an angle stands alone one letter only is often used, as A.

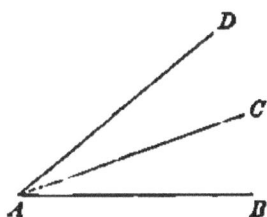

1

When one straight line standing
on another straight line makes the
adjacent angles equal to one another,
each of these angles is called a right
angle; and the straight line which
stands on the other is said to be perpendicular to it.

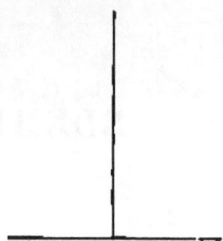

We acquire a clear idea of the magnitude of an angle by

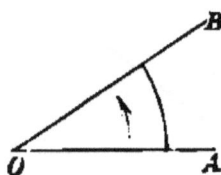

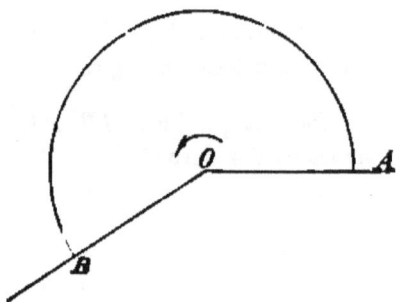

supposing it to be described by the revolution of one of the
sides OB about the end O of the other OA, which remains
fixed. When OB has revolved so far as to be in the same
line as OA but not in the same direction it has passed
over two right angles. When OB again coincides with OA,
it has passed over four right angles. The minute-hand of
a watch passes over four right angles in an hour.

All right angles are equal, and the space occupied by
two right angles is always the same; we shall often denote
it by $2R$.

An obtuse angle is greater than a right angle.

An acute angle is less than a
right angle.

THEOREM L

If any straight line CD meet another straight line AB,

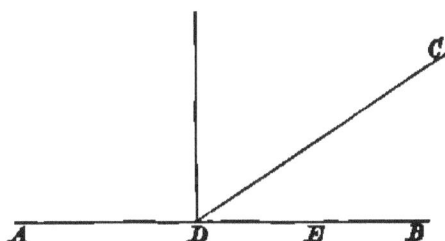

the angles CDA, CDB are together equal to two right
angles.

For they fill exactly the same space.

Conversely, if three straight lines AD, CD, BD meet in
one point D so as to make the angles CDA, CDB equal to
two right angles,

Then must AD and DB be in the same straight line.

Produce AD to E in the same straight line.

Then $CDA + CDE = 2R.$ ·

But $\qquad CDA + CDB = 2R$ (hypoth.).

Therefore $\qquad CDB = CDE.$

Therefore DE and DB must have the same direction.

Therefore DB is in the same straight line with AB.

NOTE. When two angles are together equal to two right angles, either of these angles is called the supplement of the other.

THEOREM II.

If two straight lines AB, CD cut one another, the vertical

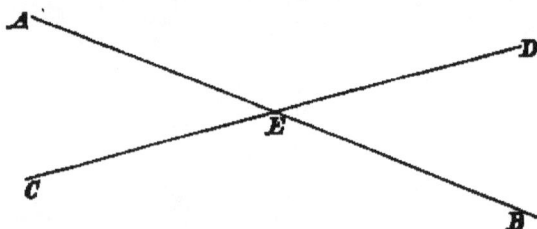

angles must be equal, viz. AEC to DEB, and AED to BEC.

For $\qquad AEC + AED = 2R,$

and also $\qquad AED + DEB = 2R.$

Therefore $\qquad AEC + AED = AED + DEB.$

Therefore $\qquad AEC = DEB.$

Similarly, $\qquad AED = BEC.$

THEOREM III.

Any straight line AB is less than the two straight lines AC, CB which have the same extremities A and B.

For AB is the shortest line that can be drawn from A to B.

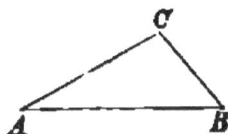

THEOREM IV.

Two straight lines AC, CB are less than two straight

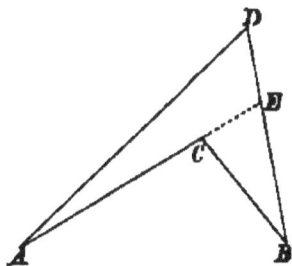

lines AD, DB which enclose them.

Produce AC to meet DB in E.

Then $CB < CE + EB$.

Therefore $AC + CB < AE + EB$.

Again, $AE < AD + DE$.

Therefore $AE + EB < AD + DB$.

Much more then $AC + CB < AD + DB$.

COR. Similarly, it may be shewn that any convex broken

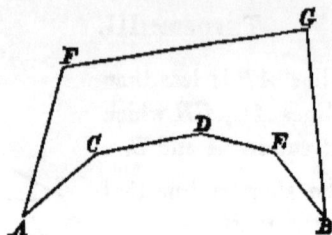

line $ACDEB$ is less than another broken line $AFGB$ by
which it is enclosed.

THEOREM V.

The perpendicular is the shortest line that can be drawn

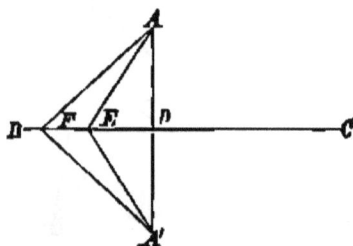

from a point A to a straight line BC; and of the others AE
which is nearer to the perpendicular is always less than one
more remote AF.

Let the figure AFD be turned about BC so that A may
fall on A'. Then, since ADB is a right angle,

$$ADB + BDA' = 2R.$$

Therefore ADA' is a straight line.

Whence $\qquad AA' < AE + EA'.$ $\hfill$ I. 3.

Therefore . $AD < AE.$

Also $AE + EA' < AF + FA'.$ I. 4.

Therefore $AE < AF.$

THEOREM VI.

· Straight lines AE, AF, equally remote from the perpendicular AD, are equal.

[AE, AF are equally remote from AD if $DE = DF$].

Let the figure ADE be turned about AD;

then since $ADE = ADF$,

DE will take the direction of DF.

And since

$DE = DF$,

the point E will coincide with
the point F.

Therefore AE will coincide
with AF, and must be equal
to it.

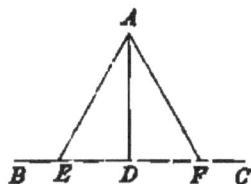

B   E   D   F   C

Cor. 1. The angle AED will coincide with the angle AFD and must be equal to it.

Cor. 2. The angle EAD will coincide with the angle FAD and must be equal to it.

Conversely, if AE, AF be equal, they must be equally remote from the perpendicular.

For (Prop. v), neither can be nearer to the perpendicular than the other.

Also, the lesser line AE is
nearer to the perpendicular than
the greater AG.

For DE can neither be
equal to (Prop. VI) nor greater
than (Prop. V), DG.

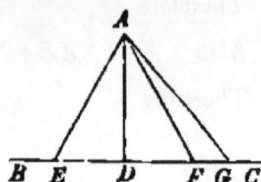

PARALLELS.

Parallel straight lines are
such as being produced ever so
far both ways do not meet.

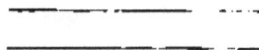

If a straight line EF cut the two straight lines AB, CD,
it makes with either of them four angles which we will denote
by (1), (2), (3), (4) ; (I), (II), (III), (IV).

If one angle of the first set
be equal to the corresponding
angle of the second set, the
other angles also will be equal
to their respective corresponding angles.

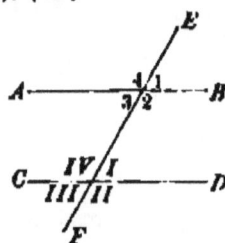

Thus, let $(1) = (I)$.

Then $(2) = (II)$, for these are the supplements of (1)
and (I).

Also (3) and $(4) = (III)$ and (IV) each to each, being ver-
tically opposite to the angles (1), (2), (I), (II).

Also $(3) = (I)$, for each is equal to (1).

Similarly $(2) = (IV)$.

Again $(I) + (2) = (1) + (2)$
$$= 2R.$$

Similarly $(3) + (IV) = 2R.$

The angles (1) and (I) are called exterior and interior angles.

The angles (3) and (I) are called alternate angles.

The angles (2) and (I) are called interior angles on the same side of EF.

THEOREM VII.

If the exterior angle EGB be equal to the interior angle GHD; then shall AB be parallel to CD.

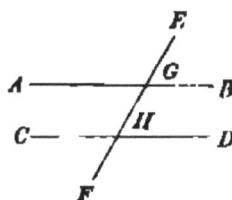

For if $\quad EGB = GHD,$

then $\quad FHC = HGA.$

Hence HC is situated with regard to GA precisely as GB with regard to HD.

Therefore if GB, HD being produced meet towards B and D, HC, GA being produced must meet towards C and A.

In which case two straight lines would enclose a space, which is absurd.

Therefore AB, CD do not meet either towards B and D, or towards C and A.

COR. 1. If either of the angles at G be equal to the corresponding angle at H, AB must be parallel to CD.

COR. 2. If the alternate angle AGH is equal to the alternate angle GHD, AB is parallel to CD.

For $\quad\quad\quad\quad AGH = EGB.$

Cor. 3. If the two interior angles BGH, GHD are together equal to two right angles, AB is parallel to CD.

For $BGH + GHD$ will then be equal to $BGH + EGB$, and therefore $EGB = GHD$.

THEOREM VIII.

If AB be parallel to CD, then shall the angle EGB be

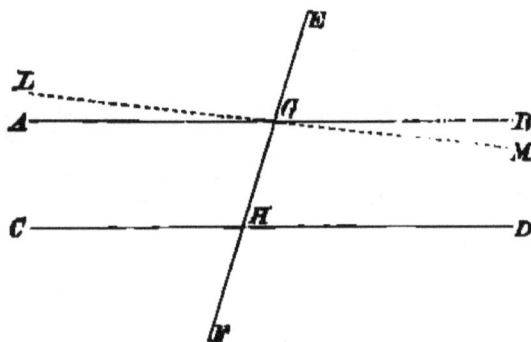

equal to the angle GHD.

For if not let EGM be equal to GHD. Then GM is parallel to HD.

Thus through the same point G two different lines are drawn each parallel to CD, which is absurd.

Cor. 1. The other angles at G are equal to the corresponding angles at H.

Cor. 2. The alternate angles AGH, GHD are equal.

Cor. 3. The two interior angles BGH, GHD are together equal to two right angles.

THEOREM IX.

Straight lines AB, CD parallel to the same line KL are parallel to one another.

Draw $EFHG$ cutting AB, KL, and CD, in F, H, and G.

Then $EFB = FHL.$

And $FHL = HGD.$

Therefore $EFB = HGD.$

Therefore AB is parallel to CD.   I. 7.

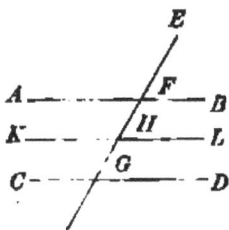

TRIANGLES.

THEOREM X.

If one side BC of a triangle ABC be produced, the exterior angle ACD is equal to the two interior and opposite angles A and B; and the three angles A, B, and C are equal to two right angles.

Through C draw CE parallel to BA.

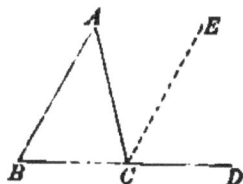

Then the alternate angle $ACE =$ the alternate angle A,

and the exterior angle $ECD =$ the interior angle B.

Therefore the whole $ACD = A + B$.

Again,

$$A + B = ACD.$$

Add to each C, then

$$A + B + C = ACD + ACB$$
$$= 2R.$$

Cor. 1. If two angles of one triangle be equal to two angles of another, the third angle of the first is equal to the third angle of the second.

Cor. 2. The exterior angle is greater than either of the interior and opposite angles.

Theorem XI.

If in any triangle ABC the side AB be equal to the side AC, then shall the angle C be equal to the angle B.

Let AD be the perpendicular drawn from A to BC.

Then since $AB = AC$, these lines are equally remote from the perpendicular.

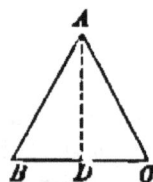

Therefore $BD = DC$,

and $C = B.$ I. 6. Cor. 1.

Cor. 1. Also $BAD = CAD$. Hence in an isosceles triangle the perpendicular bisects the base and the angle at the vertex.

Cor. 2. All the angles of an equilateral triangle are equal.

THEOREM XII.

If in any triangle ABC the side AB be greater than the side AC, then shall the angle C be greater than the angle B.

Let AD be the perpendicular. Then since AB is greater than AC, it is more remote from the perpendicular.

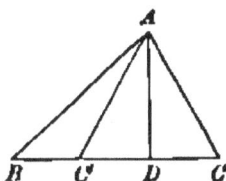

Hence if ADC be turned about AD, C will fall between D and B, as at C'.

Then the exterior angle $AC'D$ is greater than ABD; that is, $C > B$.

Conversely, if the angle C be greater than the angle B, then shall AB be greater than AC.

Cor. If the angle C be equal to the angle B, then shall AB be equal to AC.

Congruent Triangles.

Every triangle has six parts, viz. the three sides and the three angles. When the six parts of one triangle are equal to the six parts of another, each to each, the triangles are said to be congruent.

We shall show that two triangles must be congruent if in the one the following parts are equal to the corresponding parts of the other.

I.   The three sides.

II. Two sides and the included angle.

III Two angles and a corresponding side.

We shall also investigate the case of two triangles, which have two sides of the one, and an angle not the included angle, equal to the corresponding parts of the other.

THEOREM XIII.

If in the triangles ABC, $A'B'C'$,
$$AB = A'B', \quad BC = B'C', \quad CA = C'A',$$
then shall $C = C'$, $A = A'$, and $B = B'$.

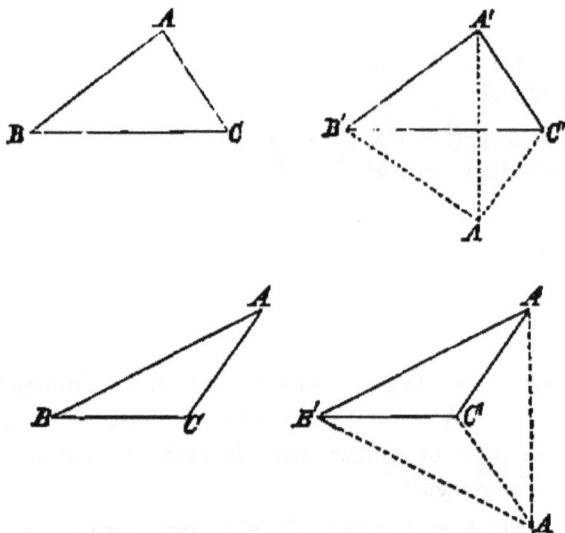

Apply the triangle ABC to the triangle $A'B'C'$ so that the point B may be on B', and BC on $B'C'$; C will coincide with C' since $BC = B'C'$. Join AA'.

Then since $B'A' = B'A$,

the angle $B'AA' = B'A'A$;   I. 11.

and since $\qquad C'A' = C'A,$

$$C'AA' = C'A'A. \qquad \text{I. 11.}$$

Therefore whole (fig. 1) or rem. (fig. 2) $B'AC' =$ whole or rem. $B'A'C'$;

i. e. $BAC = B'A'C'$;

i. e. $A = A'$.

Similarly, $\qquad B = B',$

and $\qquad C = C'.$

THEOREM XIV.

If in the triangles ABC, $A'B'C'$,

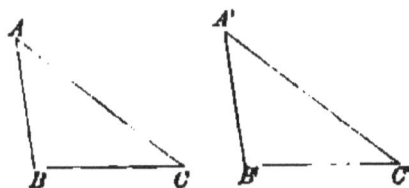

$AB = A'B'$, $BC = B'C'$, and $B = B'$,

then shall $CA = C'A'$, $C = C'$, and $A = A'$.

Apply the triangle ABC to the triangle $A'B'C'$, so that A may be on A' and AB fall on $A'B'$.

Then because $AB = A'B'$,

B will coincide with B'.

And, because the angle $B =$ the angle B',

BC will take the direction of $B'C'$.

And, because $BC = B'C'$,

the point C will coincide with the point C'.

Therefore CA must coincide with $C'A'$, and be equal to it.

And the angle C must coincide with the angle C', and be equal to it.

And the angle A must coincide with the angle A', and be equal to it.

THEOREM XV.

If in the triangles ABC, $A'B'C'$,

$$B=B', \quad C=C', \text{ and } BC=B'C',$$

then shall $A=A'$, $CA=C'A'$, and $AB=A'B'$.

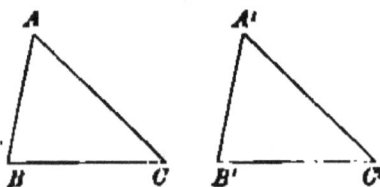

First, $A=A'$ obviously.   I. 10. Cor. 1.

To prove the rest of the proposition, apply the triangle ABC to the triangle $A'B'C'$, so that B may be on B', and BC have the direction of $B'C'$.

Then, since $BC=B'C'$,

C must coincide with C'.

Now, since $B=B'$,

BA must take the direction of $B'A'$,

and the point A must lie somewhere on $B'A'$.

Also, since $C = C'$,

CA must take the direction of $C'A'$,

and the point A must lie somewhere on $C'A'$.

Therefore A must coincide with A'.

Therefore CA must coincide with $C'A'$, and be equal to it.

And AB must coincide with $A'B'$, and be equal to it.

Ambiguous Case.

In the triangles ABC, $A'B'C'$, let

$$CA = C'A', \quad AB = A'B', \quad B = B'.$$

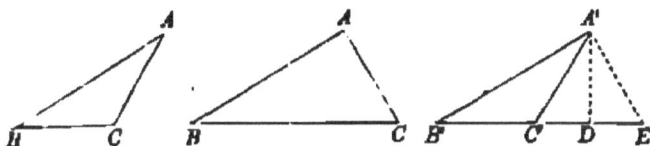

Apply the triangle ABC to the triangle $A'B'C'$, so that A may be on A', and AB have the direction of $A'B'$.

Then, since $AB = A'B'$,

the point B must coincide with B'.

And since $B = B'$,

BC must take the direction of $B'C'$.

From A' draw $A'D$ perpendicular to $B'C'$ produced, if necessary, and take $DE = DC'$.

Then because $AC = A'C'$,

these lines must be equally remote from the perpendicular $A'D$.

2

Hence AC must meet $B'C'$ either in C' or E.

If it meet $B'C'$ in C', the triangles ABC, $A'B'C'$ are congruent; otherwise not.

Now if C and C' be both acute or both obtuse angles, the points C and C' will both lie on the same side of the perpendicular,

and will coincide.

If one of the angles C and C' be acute and the other obtuse,

the triangles are not congruent.

If one of the angles as C' be a right angle, $A'C'$ will coincide with $A'D$. In this case AC must also coincide with $A'D$, since no other line equal to $A'D$ can be drawn for A' to $B'C'$.

Hence if the angles C and C' be both acute or both obtuse, or if either of them be a right angle,

the triangles are congruent.

Cor. 1. If the angle C be not equal to C', it is supplementary to it.

Cor. 2. If B and B' the angles given equal be opposite to the greater sides, the triangles must be congruent.

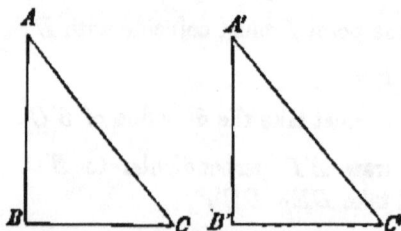

For in this case C and C' must be both acute.

Hence if the angles given equal be right angles the triangles must be congruent. This is a very common and important case, and admits easily of an independent demonstration.

For if AB, $A'B'$ be placed so as to coincide, AC, $A'C'$ being equal must be equally distant from the perpendicular. Therefore $BC = B'C'$, and the triangles are congruent.

PARALLELOGRAMS.

A Parallelogram is a four-sided figure which has its opposite sides parallel.

THEOREM XVI.

The opposite sides and angles of parallelograms are equal.

Let $ABCD$ be a parallelogram.

Then shall

$AB = CD$, $BC = DA$,

$A = C$, and $B = D$.

Join BD.

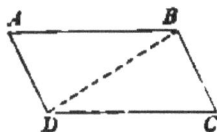

Then the alternate angle $ABD =$ the alternate angle CDB,
[I. 8. Cor. 2.

and the alternate angle $BDA =$ the alternate angle DBC.
[I. 8. Cor. 2.

Therefore in the triangles ABD, CDB,

$ABD = CDB$, $BDA = DBC$, and BD is common.

Therefore $AB = CD$, $BC = DA$, and $A = C$. [I. 15.

Also the whole angles ABC, CDA, the parts of which are separately equal, are equal.

2—2

THEOREM XVII.

The straight lines which join the extremities of equal and parallel straight lines towards the same parts are themselves equal and parallel.

Let AB be equal and parallel to CD.

Then shall BC be equal and parallel to DA.

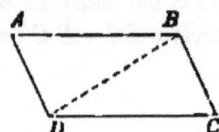

Join BD.

Then the alternate angle ABD = the alternate angle CDB.

Hence in the triangles ABD, CDB,

$AB = CD$, BD is common, and $ABD = CDB$.

Therefore $DA = BC$, and $BDA = DBC$,   [I. 14.

but BDA, DBC are alternate angles.

Therefore BC is parallel to DA.   [I. 7. Cor. 2.

THEOREM XVIII.

The exterior angles of any convex polygon are together equal to four right angles, and the interior angles to twice as many right angles as the figure has sides, diminished by four right angles.

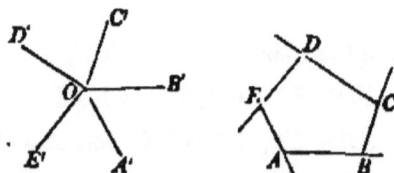

Let $ABCDE$ be any convex polygon.

Through any point O draw OB', OC', OD', OE', OA', parallel to the sides of the polygon.

Then the angle $C'OB'$ = the exterior angle at B,

and $D'OC'$ = the exterior angle at C.

Similarly the other angles at O = the other exterior angles at D, E, and A.

Therefore all the exterior angles = all the angles at O

$$= 4R.$$

Again, each interior angle with its adjacent exterior angle makes up two right angles.

Therefore all the interior and all the exterior angles are together equal to twice as many right angles as the figure has sides.

Hence the interior angles are equal to twice as many right angles as the figure has sides, diminished by four right angles.

Cor. The exterior angle of any regular polygon of n sides is equal to one n^{th} part of four right angles.

.

Loci.

Theorem XIX.

Through C the middle point of AB let CD be drawn perpendicular to AB.

Then every point on CD is equally distant from A and from B.

For if P be any point on CD, and PA, PB be joined, these lines must be equal, being equally remote from the perpendicular.

Also no point not on CD can be equally distant from A and B,

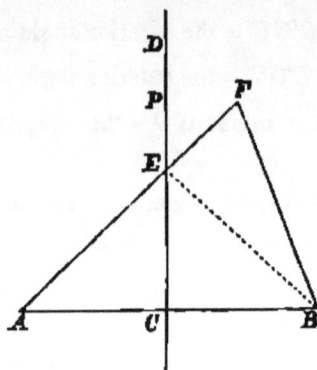

Let F be any point not on CD.

Join FA, FB; let FA cut CD in E; join EB.

Then $FB < FE + EB,$

but $EB = EA;$

therefore $FB < FA.$

Hence if we know of the point P that it must be equally distant from A and from B, we shall know that it must lie somewhere on CD.

For example, if P be the centre of any circle passing through A and B, P must be equally distant from A and B; therefore it must lie somewhere on CD.

In such a case CD is called the locus of P.

THEOREM XX.

Let AB, AC be two straight lines which meet in A, and let AD bisect the angle BAC.

Every point in AD shall be equally distant from AB, and from AC.

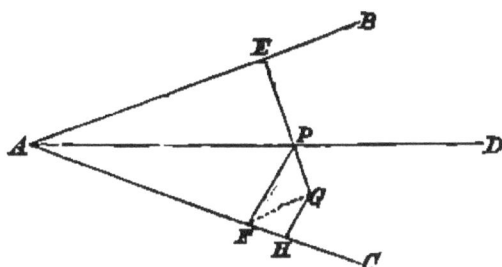

Let P be any point in AD, and draw PE, PF perpendicular to AB, and to AC.

Then in the triangles PAE, PAF,

$PAE = PAF$, $AEP = AFP$, and AP is common.

Therefore $\qquad PE = PF$. [I. 15.

Also no point not in AD can be equally distant from AB and from AC.

Let G be a point not on AD.

Let fall GE, GH perpendicular to AB, AC.

And let GE meet AD in P; let fall PF perpendicular to AC, and join GF.

Then $\qquad GH < GF$, [I. 5.

$\qquad\qquad < GP + PF$,

but $\qquad PF = PE$.

Therefore $\qquad GH < GE$.

Hence AD is the locus of a point equally distant from AB and from AC.

If AD be equally distant from AB and AC, the distance

being measured by perpendiculars let fall from a point in
AD, AB and AC are equally remote from AD, the distance
being measured along a line perpendicular to AD. For,

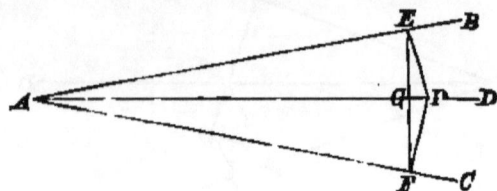

joining EF, EA, $AG = FA$, AG
each to each, and $EAG = FAG$;

therefore EAG, FAG are congruent triangles, and $EG = GF$.
Also the angles EGA, FGA are equal, therefore EF is perpendicular to AD.

The system is thus perfectly symmetrical with regard to
AD, for corresponding to every point, line, or angle on the
one side of AD, there is a like point, line, or angle on the
other.

THEOREM XXI.

The perpendiculars DG, EH, FK drawn through D, E, F,

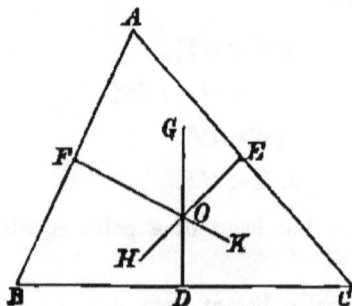

the middle points of the sides of a triangle ABC, meet in one point.

Let DG and EH intersect in O.

Then since O is a point on DG, it is equally distant from B and C.   [I. 19.

And since O is a point on EH, it is equally distant from C and A.   [I. 19.

Therefore O is equally distant from B and A.

Therefore O is a point on FK.   [I. 19.

That is, FK passes through O.

THEOREM XXII.

The straight lines AD, BE, CF, which bisect the angles of a triangle ABC, meet in one point.

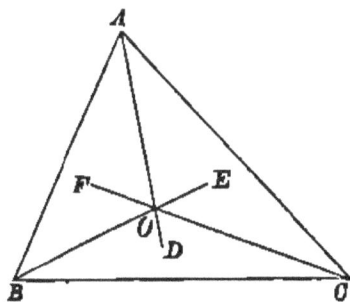

Let AD, BE intersect in O.

Then since O is a point on AD, it is equally distant from AC and AB.   [I. 20.

And since O is a point on BE, it is equally distant from BA and BC.   [I. 20.

Therefore O is equally distant from CB and CA.

Therefore it is a point on CF.   [I. 20.

That is, CF passes through O.

Theorems.

1. The straight lines which bisect two adjacent supplementary angles are perpendicular to one another.

2. If two straight lines cut one another, and straight lines be drawn bisecting the four angles which they form, these four lines will together constitute two straight lines at right angles to one another.

3. ABC is a triangle, D a point within it, shew that $AB + BC + CA$ are greater than $DA + DB + DC$, and less than twice $DA + DB + DC$.

4. The two sides of a triangle are together greater than twice the straight line drawn from the vertex to the middle point of the base.

5. If E be a point within the triangle ABC, the angle BEC is greater than the angle BAC.   I. 10.   Cor. 2.

6. ABC is a triangle, on AB produced if necessary take $AC' = AC$, and on AC take $AB' = AB$. Join $B'C'$ cutting BC in D. Show that AD bisects BAC.

7. The extremities of the base of an isosceles triangle are equally distant from the opposite sides.

8. The sum of the distances of any point on the base of an isosceles triangle from the sides is constant. What is the case if the point be taken on the base produced?

9. If the opposite sides of a quadrilateral figure be equal, it is a parallelogram.

10. If the opposite angles of a quadrilateral figure be equal, it is a parallelogram.

11. The diagonals of a parallelogram bisect one another.

12. The diagonals of a rhombus bisect one another at right angles.

13. The converse of the last two propositions.

14. If the diagonals of a parallelogram are equal, it is a rectangle.

15. If the diagonals of a parallelogram are equal and perpendicular, it is a square.

16. Straight lines which bisect the adjacent angles of a parallelogram are perpendicular to one another.

17. ABC is a right-angled triangle, C the right angle. Make $ACD = A$.

Then $BCD = B$, and $AD = BD = CD$.

18. From the angles of a square measure equal distances upon the sides in order. Join the points thus determined. The figure so formed is a square.

19. If two triangles have two sides of the one equal to two sides of the other, each to each, but the angle included by those sides in the one greater than the corresponding angle in the other, the base of the first shall be greater than that of the second.

20. The converse of 19.

21. ABC is a right-angled triangle, A the right angle: draw AD perpendicular to BC: shew that the triangles ABC, ABD are equiangular to one another. Also they have a side common, how is it that they are not congruent?

22. If through the middle point of one side of a triangle a straight line be drawn parallel to the base, it shall bisect the other side; also its length intercepted between the sides of the triangle is half that of the base.

23. Converse of 22.

24. The lines which join the middle points of the sides of any quadrilateral figure form a parallelogram: and the perimeter of this parallelogram is equal to the sum of the diagonals of the quadrilateral.

25. $ABCD$ is a parallelogram. E, F the middle points of AB, CD. Shew that BF, DE trisect the diagonal AC.

26. A pavement is to be formed of tiles of the same regular figure. Shew that the only figures which can be used are the equilateral triangle, the square, and the regular hexagon.

27. Shew that admitting two sorts of regular figures of the same sides, the pavement may be formed of

      1. Squares and octagons.

      2. Triangles and dodecagons.

28. In a given line find a point (1) equally distant from two given points, (2) equally distant from two given lines.

29. Find the locus of a point, given

    1. Its distance from a fixed point.

    2. Its distance from a fixed line.

    3. The sum of its distances from two fixed lines.

    4. The difference of its distances from two fixed lines.

30. If two sides of a triangle be produced, the lines which bisect the two exterior angles and the third interior angle meet all in one point.

31. CX, CY are two straight lines at right angles. AB a straight line of fixed length moves so that its extremities are always on CX, CY. Find the locus of the middle point of AB.

CONSTRUCTIONS.

These must not be merely indicated as in Euclid, but drawn accurately to scale with ruler and compass. In this part of the work the assistance of the master is indispensable, and we have therefore offered less explanation of our own. A good many examples of each case should be drawn, using lines and angles of known measurement, and great value should be set on accuracy and finish in the drawing. Many problems in Mechanics and Engineering can be, and practically are, solved by construction alone without calculation. We shall give some examples of this method, which is of sufficient importance to deserve serious attention.

As the circle is necessary to our constructions, we give here the ordinary definition.

A circle is a plane figure contained by one line called the circumference, and is such that all lines drawn from a point within it to the circumference are equal.

This point is called the centre of the circle.

1. To draw a straight line bisecting a given finite straight line at right angles.

Let AB be the given finite straight line.

With centre A and distance greater than half AB describe a circumference; with centre B and same distance describe a

circumference. These circumferences must intersect in two
points as at C and D. Join CD meeting AB in E; AB is
bisected at right angles in E.

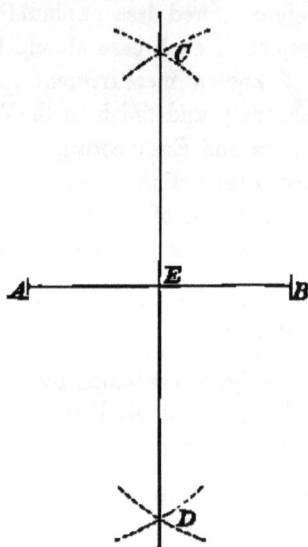

For C and D are both equidistant from A and B; therefore
both lie on the line which bisects AB at right angles. [I. 19.
Therefore CD is that line.

This is the simplest construction, but it is obviously not
essential that the radius of the arcs which intersect below
AB, should be the same as that of those which intersect
above.

2. To bisect a given angle, as A.

With centre A and any distance describe the arc BC.

With centre B and any distance describe an arc.

With centre C and the same distance describe an arc intersecting the first in D.

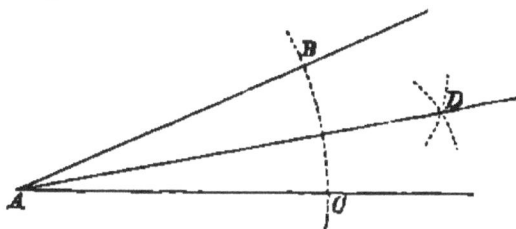

Join AD. AD bisects BAC.   [t. 13

3.   To draw a line perpendicular to a given line AB.

First. From a given point within it as C.

In CA take any point D; make $CE = CD$.

Draw FG, bisecting DE at right angles. This passes through C, and is the line required.

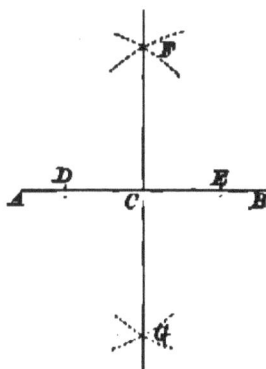

Secondly.   From a given point without it as C.

With centre C and a distance greater than the distance of C from AB describe a circumference cutting AB in D and E.

C is equally distant from D and E, and must therefore lie on the line

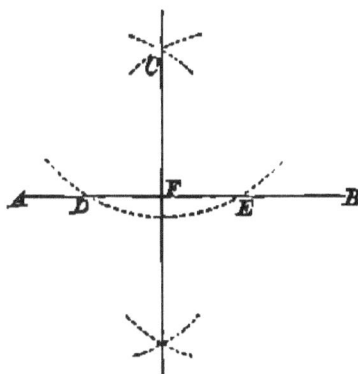

bisecting *DE* at right angles. Draw this line; it shall be the line required.

4. At a given point *B* in the straight line *BC* to make an angle equal to the given angle *A*.

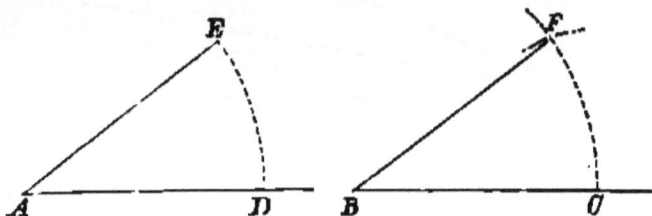

With centre *A* and any distance *AD* describe the arc *DE*. With centre *B* and the same distance describe the arc *CF*. Take off with the compass the length *DE*, and with centre *C* and this distance describe an arc intersecting *CF* in *F*. Join *BF*.

Then *CBF* is equal to *A*.   [I. 13.

For if *DE* and *CF* were joined the three sides of the triangle *EAD* would be equal to those of *FBC*, each to each.

5. Through a given point to draw a straight line parallel to a given straight line.

6. Through a given point to draw a straight line making a given angle with a given straight line.

7. (1) To construct a triangle having its sides equal to three given straight lines.

Draw a straight line *BC* equal to the greatest of the three given straight lines. With centre *B* and radius equal to the second given line describe an arc. With centre *C* and radius equal to the third given line describe an arc intersecting the former one in *A*. Join *BA, AC. ABC* is the triangle required.

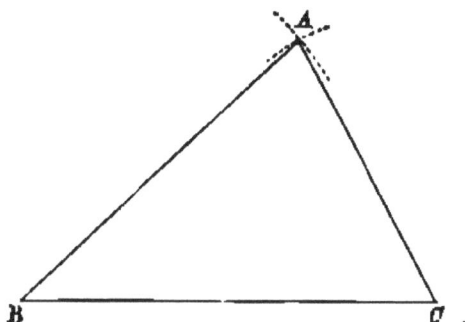

When and how would this construction fail?

To construct an isosceles triangle, or an equilateral triangle.

7. (2) To construct a triangle, having given two sides and an angle opposite to one of them.

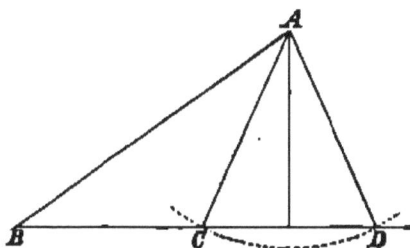

Draw a line *BC* of unlimited length, and at *B* make an angle *CBA* equal to the given angle; take *BA* equal to the side not opposite to the given angle. With centre *A* and radius

3

equal to the side opposite.the given angle describe an arc. If this meet BC in C, ABC is a triangle satisfying the given conditions.

The problem however requires further discussion.

For, first. If the side given as opposite to B be less than the perpendicular from A to BC, the arc will not meet BC, and the triangle is impossible.

Secondly. If the given side be equal to the perpendicular from A to BC, the arc will touch BC in one point, and the triangle will have a right angle.

Thirdly. If the given side be greater than the perpendicular from A to BC, the arc will meet BC in two points, as C and D.

We must now further distinguish two cases.

First. If the given side be less than AB, then C and D will both lie on the same side of B, and two triangles ABC, ABD may be constructed, both of which indifferently satisfy the given conditions.

Secondly. If the given side be greater than AB, it must therefore be more remote from the perpendicular than AB. Hence C and D must lie on different sides of B. As before, two triangles may be constructed, but one alone satisfies the given conditions, for the other does not contain the angle CBA, but its supplement.

We have supposed CBA acute; discuss the construction when CBA is obtuse.

7. (3) To construct a triangle, having given two sides and the included angle.

4. To construct a triangle, having given two angles and a side.

8. To construct a parallelogram, knowing,

    1. Two adjacent sides and a diagonal.

    2. A side and the two diagonals.

    3. The two diagonals, and the angle between them.

    4. The perimeter, a side, and any angle.

9. To trisect a right angle.

10. To construct a triangle, having given,

    1. The base, an angle at the base, and the sum or difference of the sides.

    2. The base, the difference of the sides, and the difference of the angles at the base.

    3. The base, the angle at the vertex, and the sum or difference of the sides. [7. 2].

    4. Two sides and the line drawn from the vertex to the middle point of the base.

    5. One side and two of the lines from the angles to the middle points of the opposite sides.

    6. The three lines from the angles to the middle points of the opposite sides.

It may be assumed that these lines all pass through one point which cuts off from each a third part of its length.

11. To construct a square on a given line.

12. To construct a regular hexagon on a given line.

13. To construct a regular octagon on a given line.

BOOK II.

THE radius of a circle is a straight line drawn from the centre to the circumference.

Circles which have equal radii are equal.

Any straight line drawn from one point of the circumference to another is called a chord; as AB.

The part of the circumference cut off by the chord is called an arc; it is denoted either by two letters $\overset{\frown}{AB}$ with a curved mark over them, or by three letters ACB, the first and last of which are at the extremities of the chord.

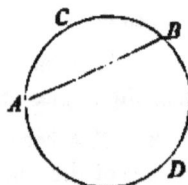

The figure ACB contained by the arc and chord is called a segment.

It will be seen that every chord divides the circumference into two arcs ACB, ADB; and the circle into two segments denoted by the same letters.

THEOREM I.

The diameter *EOC*, drawn perpendicular to a chord *AB*, bisects that chord.

Let *O* be the centre. Join *OA*, *OB*.

Then *OA*, *OB* being equal are equally distant from the perpendicular.

Therefore *AD = DB*.

Conversely, if *EOC* bisect *AB* it shall cut it at right angles.

Cor. Similarly it may be shewn that *EOC* bisects all chords parallel to *AB*.

Thus corresponding to every point, as *A*, on the circumference on one side of any diameter, there is another point, as *B*, at an equal distance on the other side.

Hence, the circumference is symmetrical with regard to any diameter, and if one semi-circumference be turned about the diameter till it fall on the other, the two will coincide throughout, each point coinciding with its corresponding point.

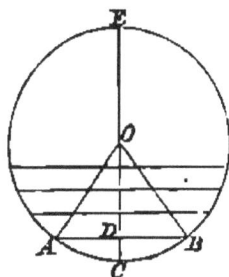

THEOREM II.

The shortest line that can be drawn from a point *A* to meet the circumference of a given circle lies on the line *AO*, which joins *A* with the centre of the circle; so also does the longest.

First case. Let A be without
the circumference BDC.

Then first, AB shall be shorter
than any other line AD.

Join OD. Then

$$AO < AD + DO,$$

but $BO = DO.$

Therefore $AB < AD.$

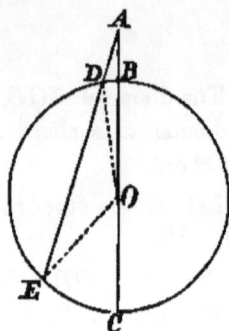

Secondly, AC is greater than any other line AE. For,
joining OE

$$AO + OE > AE,$$

but $OC = OE.$

Therefore $AC > AE.$

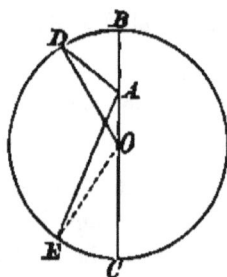

Second case. Let A be within
the circumference.

Then firstly, $OD < OA + AD$;
therefore also $OB < OA + AD.$

Therefore $AB < AD.$

Secondly, $AO + OE > AE,$
but $OE = OC.$

Therefore $AC > AE.$

Third case. Let A be upon the
circumference.

AB is now nothing.

AC is the diameter; and as before is greater than any other line
AE.

Thus the diameter is the greatest
line that can be drawn in a circle.

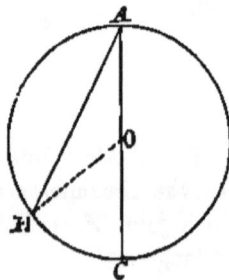

THEOREM III.

Of two straight lines AD, AE, drawn from an external point A to the circumference, AD, which is nearer to AO, is less than AE, which is more remote.

Join OD, OE, then

$$AD + DO < AE + EO \; (\text{i. 4}),$$

but $DO = EO.$

Therefore $AD < AE.$

Cor. From the same point A, two and only two equal straight lines can be drawn to the circumference, one on each side of AO.

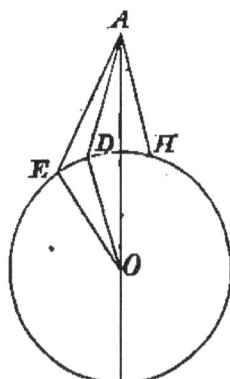

For corresponding to D, there will be another point H on the other side of AO, such that $AH = AD$, and every other line would be unequal to AH, and therefore to AD.

THEOREM IV.

Of two straight lines AD, AE, drawn from an internal point A to the circumference, AD, which is nearer to AO, is greater than AE, which is more remote.

Join OE meeting AD in F.

Then FE is the shortest line that can be drawn from F to the circumference (ii. 2).

Therefore $FD > FE$:

add to each AF.

Then $AD > AF + FE$,

but $AF + FE > AE$.

Therefore $AD > AE$.

Cor. 1. From the same point A, two and only two equal straight lines can be drawn to the circumference, one on each side of AO.

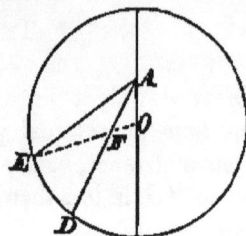

Cor. 2. The same proof will hold good if A be taken on the circumference.

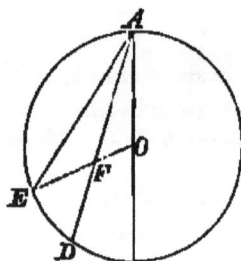

Intersection and Contact.

THEOREM V.

A straight line xy cannot cut the circumference in more than two points. (Fig. on next page.)

For only two equal straight lines can be drawn from the centre O to xy.   [I. 6.

A straight line which cuts a circle is called a secant; a straight line which meets the circumference but does not cut it is called a tangent: as DBE.

A straight line drawn through the point of contact perpendicular to the tangent is called a normal.

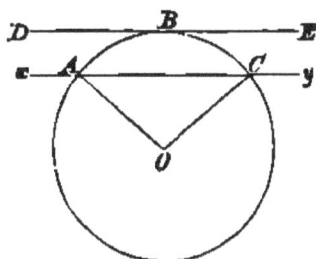

We shall find that the normal always passes through the centre.

If one circumference meets another but does not cut it, it is said to touch it.

THEOREM VI.

The straight line AB drawn at right angles to the radius OA from its extremity A is a tangent to the circle at A.

For the perpendicular OA is shorter than any other line OC that can be drawn from O to A.

Hence every other point in AB except A lies without the circumference.

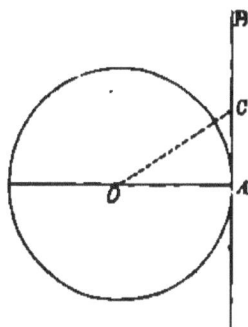

Conversely. Every tangent is perpendicular to the radius drawn to the point of contact.

For the radius OA is the shortest line that can be drawn to AB; it is therefore perpendicular to it.

COR. 1. There is but one tangent at any point.

COR. 2. The normal at every point passes through the centre.

THEOREM VII.

The circumferences of two circles O and I cannot cut one another in more than two points; nor touch in more than one.

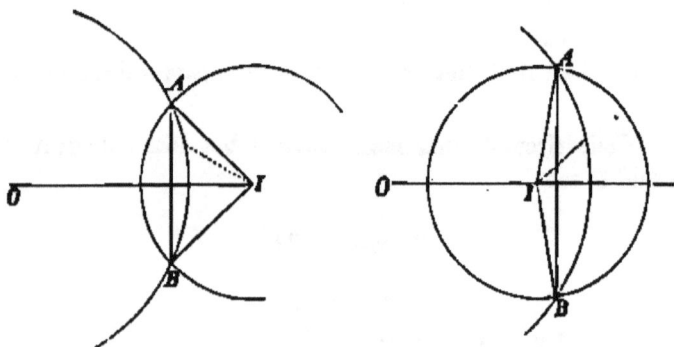

For every point on the circumference of I is equally distant from the centre I.

But only two equal straight lines can be drawn from I to the circumference of O. [II. 3. Cor.

Therefore the circumference of O cannot meet that of I in more than two points.

Let A and B be points where the circumference of O meets that of I. Join IA, IB.

Then since $IA = IB$ these lines lie on different sides of the shortest line that can be drawn to the circumference of O, and either is greater than that line. [II. 3. Cor. II. 2.

Hence there is at least one point on the circumference of O nearer to I than A or B, and therefore within the circle I.

Therefore if the circumference of O meets that of I in two points it must cut it;

That is it cannot touch it in more than one point.

THEOREM VIII.

The straight line AB which joins the points of intersection A, B, is bisected at right angles by OI, which joins the centres O and I. (Fig., Theorem VII.)

For if AB be bisected at right angles the centre of either circle must lie on the bisecting line. [I. 19.

THEOREM IX.

If two circles touch each other they have a common tangent at the point of contact.

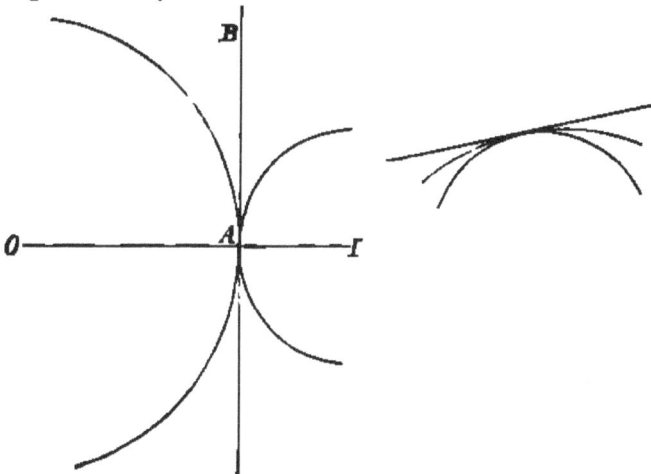

If one circle touch the other internally the tangent to the outer circle is obviously the tangent to the inner.

If the circles touch externally let AB be the tangent at A to O.

Then OAB is a right angle.

And since OA is the shortest line that can be drawn to the circumference of I, OA produced must pass through the centre I.

Therefore IA is a radius, and IAB a right angle.

Therefore AB is the tangent at A to I.

COR. If two circles touch one another, the straight line joining their centres must pass through the point of contact.

For the centre of either circle must lie on the normal at the point of contact.

Arcs and Angles.

THEOREM X.

In equal circles or in the same circle, equal arcs subtend equal angles at the centre, and also equal chords.

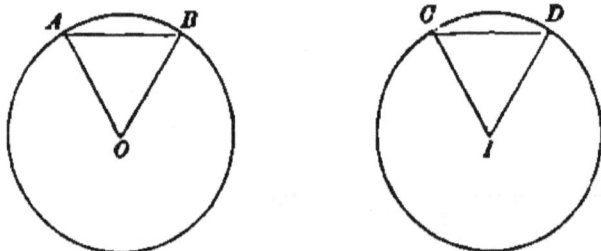

Let O and I be equal circles, and let the arc AB be equal to the arc CD.

Then shall $\qquad AOB = CID$,

and $\qquad$ the chord $AB =$ the chord CD.

Apply the circle O to the circle I so that the point O may be on I. The circumferences will coincide throughout.

Let the circle O be turned around its centre till A coincide with C. Then since the arc AB = arc CD,

B must coincide with D.

Therefore BO must coincide with DI,

and AOB must coincide with CID.

Therefore $AOB = CID$.

Also AB must coincide with CD.

Therefore $AB = CD$.

COR. 1. It is also evident that if the arc AB were greater than the arc CD the angle AOB would be greater than CID; and if less, less.

COR. 2. If the arc AB were greater than the arc CD, the chord AB would be nearer than CD to the diameter through A, and would therefore be greater than CD; and if the arc AB were less than the arc CD, the chord AB would be less than the chord CD. [II. 4. Cor. 2.

Conversely; if AOB be greater than, equal to, or less than CID, the arc AB is greater than, equal to, or less than the arc CD; or if the chord AB be greater than, equal to, or less than the chord CD, the arc AB is greater than, equal to, or less than the arc CD.

THEOREM XI.

In equal circles or in the same circle angles at the centre are to one another as the arcs on which they stand.

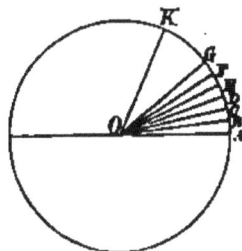

Let the circumference of the circle O be divided into a number of equal arcs AB, BC, &c. Draw the radii. Then since the arcs are equal, the angles at the centre are likewise all equal.

Hence $\widehat{AC}$ being twice as great as $\widehat{AB}$, AOC is twice as great as AOB.

And $\widehat{AD}$ being three times as great as $\widehat{AB}$, AOD is three times as great as AOB.

And if $\widehat{AK}$ be m times as great as $\widehat{AB}$, AOK is m times as great as AOB.

Similarly if $\widehat{AG}$ be n times as great as $\widehat{AB}$, AOG is n times as great as AOB.

Therefore $\widehat{AK} : \widehat{AG} = m : n.$

And $AOK : AOG = m : n.$

Therefore $\widehat{AK} : \widehat{AG} = AOK : AOG.$

On this property of the circle is based the ordinary method of measuring angles. The circumference of a circle is usually divided into 360 parts. The angle subtended by one of these parts is called a degree. Instruments for the measurement of angles are furnished with an arc of a circle on which these divisions are marked, and an angle is measured by ascertaining by how many of these divisions it is subtended at the centre of the circle. If the number be m the angle is m times as great as an angle of one degree, and is called an angle of m degrees. Degrees are subdivided into minutes and seconds.

THEOREM XII.

In equal circles, or in the same circle, equal straight lines are equally distant from the centre.

Let O and I be equal circles, AB and CD equal straight lines in them, these shall be equally distant from the centres.

Place O upon I and let it be turned round till the point A coincide with the point C.

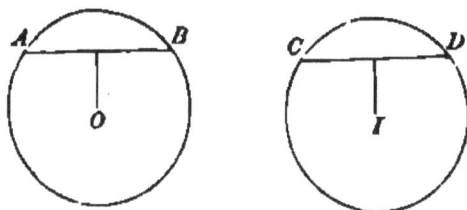

Since the chord $AB =$ the chord CD,

the arc $AB =$ the arc CD.   [II. 10. Cor. 2.

Therefore the point B will coincide with D, and the chord AB with the chord CD.

Therefore the perpendiculars upon these lines from the centres will coincide with and be equal to one another.

COR. Also if AB be greater than CD, the arc AB must be greater than the arc CD, and when the circle O is placed as above on I the point B must fall beyond D. AB is then obviously nearer to the centre than CD.

Conversely. Straight lines which are equally distant from the centre are equal; also a line which is nearer to the centre is greater than one more remote.

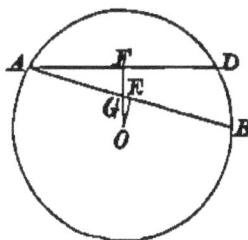

THEOREM XIII.

The angle at the centre is double of the angle at the circumference which stands upon the same arc.

Let AOB, ACB be angles at the centre and at the circumference which stand upon the same arc AB.

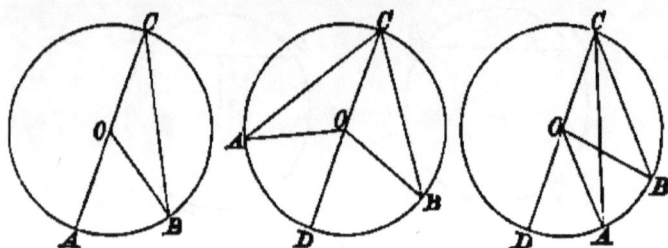

Then shall $\qquad AOB = 2ACB.$

First, let AC, one of the lines, containing ACB, pass through O.

Then $\qquad\qquad OC = OB.$

Therefore $\qquad\qquad B = C,$

and $\qquad\qquad B + C = 2C.$

But the exterior angle $AOB = B + C$.

Therefore $\qquad\qquad AOB = 2ACB.$

Secondly, let neither AC nor BC pass through O. Join CO and produce it to D.

Then $\qquad\qquad DOB = 2DCB.$

and $\qquad\qquad DOA = 2DCA.$

Therefore whole (fig. 2) $AOB =$ twice whole ACB.

or $\qquad$ rem. (fig. 3) $AOB =$ twice rem. ACB.

When the arc ADB is greater than a semicircle, the angle AOB will be greater than two right angles.

We shall call it the convex angle AOB, and denote

it by AOB, to distinguish it from the angle AOB standing on the arc ACB.

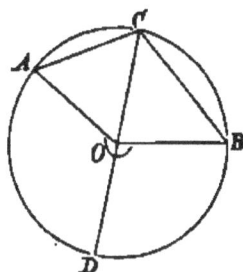

It is easily seen that every part of the proof given above applies directly to the case of a convex angle such as AOB.

THEOREM XIV.

Angles in the same segment, or which stand upon the same arc, are equal.

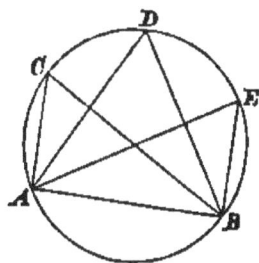

For each is half of the angle at the centre.

Thus $ACB = ADB = AEB$.

Theorem XV.

If the whole circumference be divided into two parts, ACB, ADB, the angles which stand upon these parts are supplementary; that is,

the angle ADB + the angle ACB

$$= 2R.$$

Find the centre O and join AO, BO.

Then $ADB = \frac{1}{2} AOB$,

and $ACB = \frac{1}{2}$ the convex angle AOB.

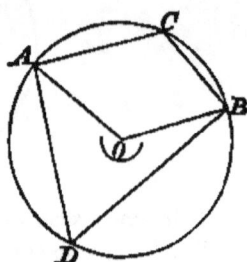

Therefore $ADB + ACB = \frac{1}{2} \{AOB + AOB\}$,

but $AOB + AOB = 4R$.

Therefore $ADB + ACB = 2R$.

Cor. 1. The opposite angles of any quadrilateral figure inscribed in a circle are equal to $2R$.

Cor. 2. The angle on a semicircumference or in a semicircle is a right angle.

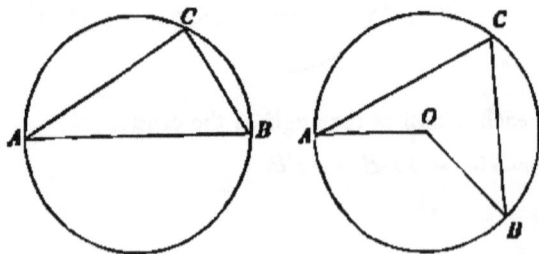

Cor. 3. The angle on an arc less than a semicircumference is less than a right angle; and the angle on an arc greater than a semicircumference is greater than a right angle.

Theorem XVI.

If a straight line EF touch a circle in the point B, and from B BA be drawn cutting the circle, the angle ABE shall be equal to the angle upon the arc ADB, and the angle ABF to that upon the arc ACB.

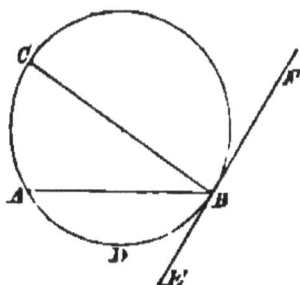

Draw the normal BC, which will pass through the centre and divide the circumference into two equal parts. Then since the arc $CA +$ arc $ADB = $ a semicircumference,

the angle on $\overset{\frown}{CA} +$ the angle on $\overset{\frown}{ADB} = R$;   [II. 15. Cor. 2.

but $CBA + ABE = R,$

therefore $CBA + ABE =$ the angle on $\overset{\frown}{CA} +$ the angle on $\overset{\frown}{ADB},$

and CBA is the angle on $\overset{\frown}{CA}.$

Therefore $ABE =$ the angle on $\overset{\frown}{ADB}.$

Therefore also the supplements of these are equal, viz.

the angle ABF and the angle on $\overset{\frown}{ACB}.$

4—2

We give another proof.

Let ACB, AHB be angles on the arc ADB.

Produce HB to K.

Then $ABH, BHA, HAB = 2R$ [ɪ. 10.
                $= ABE + ABF.$

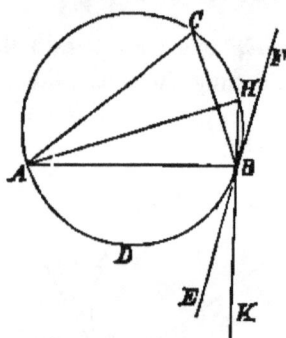

Take away the common ABH.

Then $BHA + HAB = ABE + HBF,$

therefore $ACB + HAB = ABE + HBF.$ [ɪɪ. 14.

Now let H be moved along the circumference up to B, AH, HB remaining always joined.

Then HAB is continually diminished,

also HBF is continually diminished,

and HK continually approaches the position of EF.

And when H coincides with B,

                $HAB = 0,$ and $HBF = 0.$

Therefore $ACB = ABE.$

1. Determine the locus of a point which is always equally distant from a given circumference. Discuss the several cases.

2. A line of fixed length remains always parallel to itself while one extremity describes a circle, what is the locus of the other extremity?

3. If a straight line cut two concentric circles, the parts intercepted between the circumferences are equal.

4. Through any point within a circumference chords are drawn equally inclined to the diameter through that point, shew that they are equal.

5. The same if the point be without the circumference.

6. Find the longest and shortest lines that can be drawn between two given circumferences.

7. If two equal circumferences intersect at right angles, the common chord is equal to the distance of the centres. Circumferences are said to intersect at right angles when the tangents at the point of section are at right angles.

8. Under what conditions will one circumference cut, touch, or enclose another?

9. A circle being given, how many circles of the same radius will enclose it?

10. Through a given internal point to draw the shortest possible chord.

11. If two circumferences intersect and parallel chords be drawn through the points of section, these chords are equal.

12. If two chords intersect within a circle, the angle which is contained between them is equal to half the sum of the angles subtended at the centre by the arcs which they intercept.

13. If the two chords intersect without the circle, to what is the angle contained between them equal?

14. ABC is a triangle, and B is greater than C, on AC take $AD = AB$ and join BD. Shew that ADB is equal to half the sum of B and C, and DBC to half their difference.

15. The lines joining the extremities of two diameters are parallel.

16. If triangles be formed by joining the extremities of intersecting chords, these triangles are equiangular to one another.

The chords may intersect either within or without the circle, and their extremities may be joined in different ways.

17. If parallel straight lines meet a circumference, they cut off equal arcs.

18. What is the locus of a point from which a given line is always seen under a constant angle?

19. ABC is a triangle, AD, BE, CF perpendiculars let fall on the opposite sides. Shew that these bisect the angles of the triangle DEF.

20. If two circles touch each other and secants be drawn through the point of contact, the lines joining their extremities are parallel.

21. AB is the diameter of a circle, AC, any chord, is produced to M so that CM is equal to CB. What is the locus of M?

22. Chords are drawn through a fixed point, find the locus of the middle points of these.

23. A line is drawn from a fixed point to a circumference, find the locus of its middle point.

24. Shew that a circle may be described about any regular polygon; i.e. any polygon having equal sides and angles.

25. Can a circle be described about any quadrilateral figure?

CONSTRUCTIONS.

1. Three points being given, find the centre of the circle which passes through them.   [L 19. 21.
When does the construction fail?

2. The circumference of a circle being given, find its centre.

3. Construct a circle, passing through two points, and having
    1. Its centre on a given straight line.
    2. Its centre on a given circumference.
    3. Its radius equal to a given straight line.

4. From a given point to draw tangents to a given circle.

The point being without the circumference, draw a line joining it with the centre; on this line as diameter describe a circle, which will cut the given circumference in two points. The lines joining these with the given point are the tangents required. Prove, that these lines are tangents, that they are equal, that they subtend equal angles at the centre, and that they are equally inclined to the diameter passing through the external point.

If the given point be on the circumference?

5. Construct three circumferences of equal radius so as to touch one another, and draw another circumference touching all three. Two cases.

6. Bisect a given arc.

7. From a given circle to cut off a segment containing an angle equal to a given angle.   [II. 16.

8.　On a given line to construct a segment of a circle containing an angle equal to a given angle.　　　　[II. 16.

9.　From a given point to draw a secant to a given circumference, so that the chord intercepted may have a given length.　　　　[II. 12.

10.　To construct a right-angled triangle, knowing,

　　1.　The hypothenuse and one of the acute angles.

　　2.　The hypothenuse and a side.

11.　To construct an isosceles triangle knowing the base, and the angle at the vertex.　　　　[II. 14.

12.　To construct a square knowing the diagonal.

13.　To construct a circle of given radius,

　　1.　Passing through a given point, and touching a given line, or a given circumference.

　　2.　Touching two lines.

　　3.　Touching a straight line and a circumference.

　　4.　Touching two circumferences.

14.　To draw a common tangent to two given circumferences. How many such tangents can be drawn?

A circle is said to be inscribed in a rectilineal figure when the circumference touches every side of the figure, and to be described about such a figure when the circumference passes through all the angular points of the figure.

15.　To describe a circle about a given triangle.

16.　To describe a circle about a given square.

17.　To inscribe a circle in a given triangle.　　　[I. 20. 22.

18.　To construct a circle touching three given straight lines.

The three straight lines are supposed to meet so as to form a triangle, and there are four circles satisfying the conditions. Shew that the lines which join the centres of the exterior circles pass through the angular points of the triangle; also that the tangents drawn from the angular points to the further exterior circles are all equal.

19. In a given circle inscribe,

    1. An equilateral triangle.   [II. 16.

    2. A square.   [II. 15. Cor. 2.

    3. A regular hexagon.

20. Construct a circle touching a given line in a given point, and

    1. Passing through a given point.

    2. Having a given radius.

    3. Touching another given line.

BOOK III.

MANY geometrical questions are most simply treated by the ordinary methods of arithmetic.

Of this kind are all questions of ratio and proportion, and most questions as to the area of figures.

It is therefore most important to acquire clear ideas as to the way in which the different geometrical magnitudes are measured.

A straight line is measured by ascertaining how many yards, feet, inches, or other units of length it contains.

An angle is measured by ascertaining how many degrees, minutes, seconds, or other units of angular space it contains.

A surface is measured by ascertaining how many acres, square yards, square feet, or other units of surface it contains.

The measurement of straight lines is familiar to everybody.

The measurement of angles we have already explained.

The measurement of the area or surface of figures is less easy; but the measurement of a rectangle is a simple operation, and we shall shew that the measurement of any rectilineal figure may eventually be reduced to the measurement of a rectangle.

In all that follows, whenever the product of two magnitudes is spoken of, it must be understood of the product of the numbers representing these magnitudes.

A line cannot be multiplied by a line, nor a surface by a surface, but the numbers representing a line or a surface may be combined in any way, in accordance with the ordinary laws of arithmetic.

We shall often have occasion to employ the simpler laws of numerical proportion, and it may be convenient to state the more important of them here. They are these.

If
$$\frac{a}{b} = \frac{c}{d},$$

Then,

I.
$$ad = bc,$$

II.
$$\frac{a}{c} = \frac{b}{d},$$

III.
$$\frac{a+b}{b} = \frac{c+d}{d},$$

IV.
$$\frac{a-b}{b} = \frac{c-d}{d}.$$

Also if
$$\frac{a}{b} = \frac{b}{c},$$

Then
$$\frac{a}{c} = \frac{a^2}{b^2} = \frac{b^2}{c^2}.$$

In applying arithmetic to geometry we are met by the preliminary difficulty that certain magnitudes are incommensurable with regard to one another, that is, cannot be exactly

measured with the same unit. This difficulty is less serious
than it appears, and we shall pass it by, giving proofs which
are directly applicable only to commensurable magnitudes.
The more advanced student will easily convince himself that
whatever is true of these is true also of magnitudes which are
not commensurable.

EQUAL FIGURES.

The word equal and the sign ▬ when used of figures
denote equality of area, and nothing more.

Congruent figures are equal, since they can be made to
coincide, but equal figures are not necessarily congruent.
Two fields may be of the same size, though their boundaries
differ in every respect.

THEOREM L.

A parallelogram $ABCD$ is bisected by its diagonal.

For the triangles ABD, BDC have been shown to be congruent (I. 16), and are therefore equal.

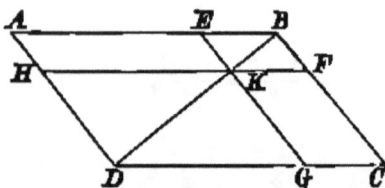

Cor. The complements about the diameter, that is, the
figures AK, KC are equal.

For the triangle ABD ▬ the triangle CDB;

also   the triangle EBK ▬ the triangle FKB,

and   the triangle HKD = the triangle GDK;

Therefore the remaining figure AK ▬ the remaining
figure KC.

THEOREM II.

Parallelograms upon the same base and between the same parallels are equal.

Let $ABCD$, $EBCF$ be parallelograms upon the same base BC and between the same parallels BC and AF, they shall be equal one to another.

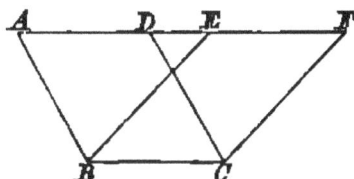

For $AD = BC$ and $BC = EF$.   [I. 16.

Therefore $AD = EF$.

Therefore whole $AE =$ whole DF.

Whence in the triangles EAB, FDC

$EA = FD$, $AB = DC$ and $BE = CF$.

Therefore the triangles are congruent and equal.

From the whole fig. $ADCF$ take away the triangle FDC, there is left the parallelogram $ABCD$.

Again, from the whole fig. $ABCF$ take away the triangle EAB.

There is left the parallelogram $EBCF$.

Therefore the parallelogram $ABCD =$ the parallelogram $EBCF$.

COR. Parallelograms upon equal bases and between the same parallels are equal.

For they may always be placed so that their equal bases coincide, which will be equivalent to their being on the same base.

THEOREM III.

Any triangle ABC is half of a parallelogram $DBCE$ on the same base and between the same parallels.

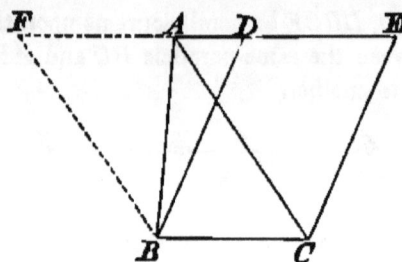

Complete the parallelogram $ACBF$.

Then $ACBF = DBCE$.

But ABC is half of $ACBF$; therefore of $DBCE$.

THEOREM IV.

Triangles upon the same base and between the same parallels are equal.

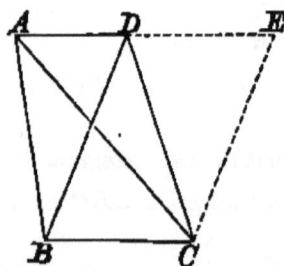

For each is half of any parallelogram on that base and between those parallels.

COR. 1. Triangles upon equal bases and between the same parallels are equal.

Conversely, equal triangles upon the same base or upon equal bases in the same straight line are between the same parallels.

MEASUREMENT OF FIGURES.

DEF. Every right-angled parallelogram is called a rectangle.

The altitude of a parallelogram is the perpendicular distance of the opposite side from the base.

The altitude of a triangle is the perpendicular distance of the vertex from the base.

THEOREM V.

The measure of a rectangle is the product of its base multiplied by its altitude, the square on the unit of length being taken as the unit of surface.

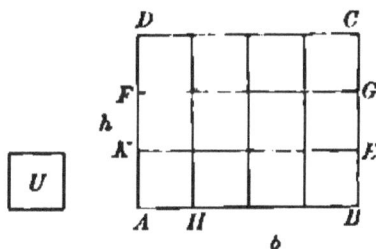

Let $ABCD$ be a rectangle.

Divide AB, AD into parts each equal to the unit of length, and through the points of division draw straight lines parallel to AB and AD.

The figures so formed are all squares, and all equal to the square KH.   [III. 2.

Hence if AB contain b equal parts, KB will contain b squares each equal to KH.

Therefore $KB = KH \times b.$

Again the figures DG, FE are equal to each other and to KB. [III. 2.

Hence if AD contain h equal parts, DB will contain h figures each equal to KB.

Therefore $DB = KB \times h,$

but $KB = KH \times b.$

Therefore $DB = KH \times b \times h.$

Hence if KH be taken for the unit of surface,

the measure of $DB = b \times h,$

where b and h are the numbers which express the lengths of AB and AD.

COR. The measure of a square is the square of the measure of the base.

THEOREM VI.

The measure of a parallelogram is the product of its base multiplied by its altitude.

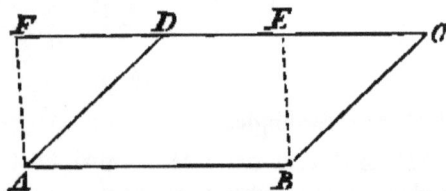

Let $ABCD$ be a parallelogram, draw BE perpendicular to AB, and complete the parallelogram $ABEF$.

Then the parallelogram $ABCD =$ the rectangle $ABEF$.

Therefore its measure $= AB \times BE$.

Cor. 1. Parallelograms of the same or equal altitude are to one another as their bases.

Let *ABCD, EFGH* be parallelograms of equal altitude.

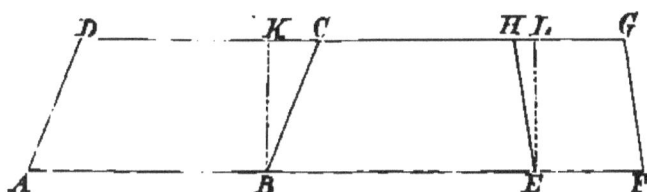

Draw the perpendiculars *BK, EL.*

Then $\dfrac{\text{parallelogram } DB}{\text{parallelogram } HF} = \dfrac{AB \times BK}{EF \times EL} = \dfrac{AB}{EF}.$

Cor. 2. Parallelograms upon the same or equal bases are to one another as their altitudes.

THEOREM VII.

The measure of a triangle is half the product of its base multiplied by its altitude.

Cor. 1. Triangles of the same or equal altitude are to one another as their bases.

Cor. 2. Triangles on the same or equal bases are to one another as their altitudes.

Def. A four-sided figure two of whose sides are parallel is called a trapezium.

THEOREM VIII.

The measure of a trapezium is the product of the altitude and half the sum of the parallel sides.

Let *ABCD* be a trapezium, of which *AB, CD,* are the parallel sides.

5

Draw the diagonal *BD*, and through *A* draw *AE* parallel to *BD*, cutting *CD* produced in *E*. Join *BE*.

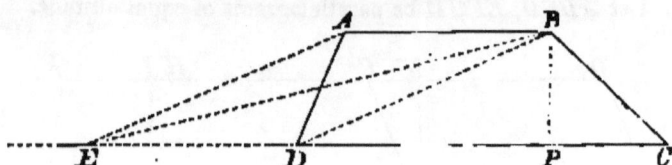

Then the triangle ABD = triangle EBD,   [III. 4. because they are on the same base and between the same parallels.

Add to each the triangle *BCD*.

Then the trapezium $ABCD$ = the triangle BEC.

But the measure of the triangle BEC is $\frac{1}{2} EC \times BP$ (*BP* being the altitude).

And $EC = ED + DC = AB + CD$ since $ABDE$ is a parallelogram.

Therefore the measure of the trapezium

$$= \frac{1}{2}(AB + CD) \times BP.$$

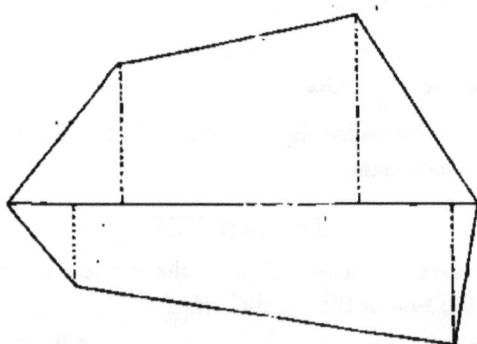

This proposition may be applied to obtain the area of a

convex polygon. The figure shews the necessary construction, by which the polygon is divided into triangles and rectangular trapeziums.

THEOREM IX.

The square on the sum AC of two straight lines AB, BC is equal to the squares on AB and on BC, together with twice the rectangle AB, BC.

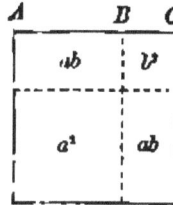

Let a be the measure of AB,

b .... BC.

Then $a + b$ is .... AC.

Hence

The square on $AC = (a + b)^2$ [III. 5. Cor.

$$= a^2 + 2ab + b^2$$

= square on AB + square on BC

+ 2 rectangle $AB . BC$.

THEOREM X.

The square on the difference AC of two straight lines AB, BC is equal to the squares on AB, BC diminished by twice the rectangle $AB . BC$.

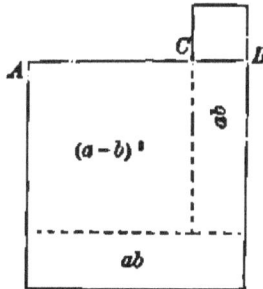

Let $AB = a$, $BC = b$.

Then $AC = (a - b)$.

And the square on $AC = (a - b)^2$

$$= a^2 - 2ab + b^2$$

= square on AB + square on BC

− 2 rectangle $AB . BC$.

5—2

THEOREM XI.

The rectangle contained by the sum of two lines and their difference is equal to the difference of their squares.

Let $AB = a$, $BC = b$.

Then $AB + BC = a + b$,

$AB - BC = a - b$.

Therefore the rectangle contained by $AB + BC$, and $AB - BC$

$= (a + b)(a - b)$

$= a^2 - b^2$

$=$ square on $AB -$ square on BC.

Produce AB to A', making $BA' = BA$.

Then $AC = AB - BC$, and $A'C = AB + BC$, therefore the rectangle $AC . A'C =$ square on $AB -$ square on BC.

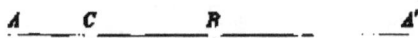

Again, produce AB to C' making $BC' = BC$.

Then $AC = AB - BC$, and $AC' = AB + BC$, therefore the rectangle $AC . AC' =$ the square on $AB -$ square on BC.

These forms of the theorem are often employed.

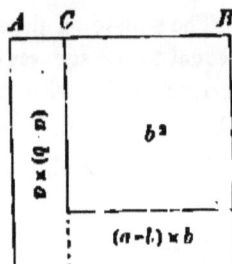

THEOREM XII.

In any right-angled triangle, the square which is described on the side subtending the right angle is equal to the squares described on the sides which contain the right angle.

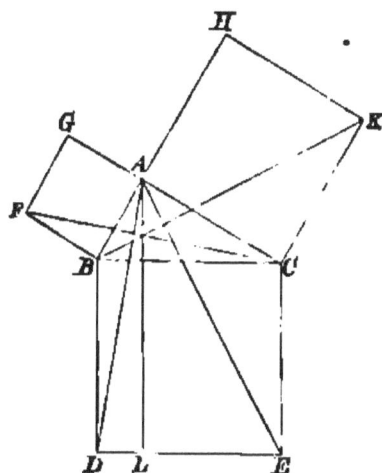

Let ABC be a right-angled triangle, having the right angle BAC: the square described on the side BC shall be equal to the squares described on the sides BA, AC.

On BC describe the square $BDEC$, and on BA, AC describe the squares GB, HC: through A draw AL parallel to BD or CE: and join AD, FC.

Then, because the angle BAC is a right angle,

and that the angle BAG is also a right angle,

the two straight lines AC, AG on the opposite sides of AB, make with it at the point A the adjacent angles equal to two right angles:

Therefore CA is in the same straight line with AG.

For the same reason, AB and AH are in the same straight line.

Now the angle DBC is equal to the angle FBA, for each of them is a right angle.

Add to each the angle ABC.

Therefore the whole angle DBA is equal to the whole angle FBC.

And because the two sides AB, BD are equal to the two sides FB, BC each to each;

and the angle DBA is equal to the angle FBC,

therefore the triangle ABD is equal to the triangle FBC.

Now the parallelogram BL is double of the triangle ABD, because they are on the same base BD and between the same parallels BD, AL.

And the square GB is double of the triangle FBC, because they are on the same base FB and between the same parallels FB, GC.

But the doubles of equals are equal to one another.

Therefore the parallelogram BL is equal to the square GB.

In the same manner by joining AE, BK, it can be shewn that the parallelogram CL is equal to the square CH.

Therefore the whole square $BDEC$ is equal to the two squares GB, HC.

And the square $BDEC$ is described on BC, and the squares GB, HC on BA, AC.

Therefore the square described on the side *BC* is equal to the squares described on the sides *BA*, *AC.*

Wherefore, in any right-angled triangle, &c. Q. E. D.

The preceding proof is that given by Euclid.

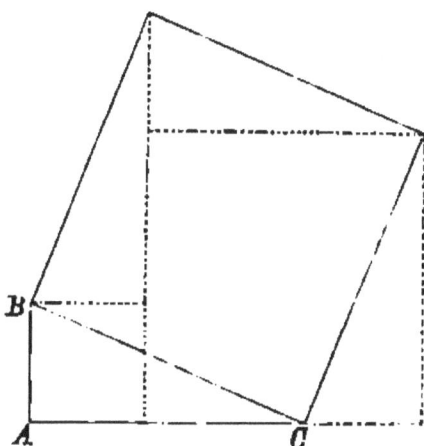

The proposition may also be demonstrated by showing that the square on the hypothenuse may be divided into parts capable of being so placed as to coincide with the squares on the sides.

The figure shews the necessary construction.

We shall give further on a full discussion of this important proposition.

THEOREMS.

1. THE diagonals divide a parallelogram into four equal triangles.

2. If two triangles have two sides of the one equal to

two sides of the other, each to each, and the angles contained by these sides supplementary, the triangles are equal.

3. If the middle points of the sides of a triangle be joined, the area of the triangle thus formed is one quarter of the area of the original triangle.

4. Every line which passes through the intersection of the diagonals of a parallelogram, divides the figure into two equal parts.

5. If a point within a parallelogram be joined to the vertices, the two triangles formed by the joining lines and two opposite sides are together equal to half the parallelogram. What is the case when the point is outside?

6. $ABCD$ is a parallelogram; through E, F, &c. on AB, EH, FK, &c. are drawn parallel to BC: shew that the triangles AHE, EKF, &c. are together equal to half the parallelogram.

7. $ABCD$ is a parallelogram, P a point outside it: the triangles PAB and PAD are together equal to the triangle PAC.

What is the case if the point be inside?

8. The measure of a triangle is equal to half the product of the perimeter and the radius of the inscribed circle.

9. The area of a trapezium is equal to twice the area of the triangle formed by joining the extremities of one non-parallel side to the middle point of the other.

10. Being given the base and height of a trapezium of known area, to find the areas of the two triangles of which it is the difference.

11. The sum of the perpendiculars let fall on the sides of an equilateral triangle from any point within it is constant.

12. If a chord of a circle be divided in any point, the rectangle contained by the segments is equal to the difference of the squares on the radius, and on the line drawn from the centre to the point.

13. $ABCD$ is a rectangle, P any point; the squares on PA and PC are equal to the squares on PB and PD.

14. The squares on the diagonals of a parallelogram are together equal to the squares on the four sides.

15. In any triangle the squares on the sides are together equal to twice the square on half the base together with twice the square on the line drawn from the vertex to the middle point of the base.

16. The squares of the diagonals of a quadrilateral are together less than the squares on the four sides by four times the square on the line joining the middle points of the diagonals.

17. If two sides of a triangle are given the area is greatest when they contain a right angle.

18. Of equal triangles on the same base the isosceles has the least perimeter.

Let BAC, BEC be equal triangles on the same base BC, and therefore between the same parallels. Let BAC be isosceles. Produce BA to D, making $AD = BA$ or AC. Join ED. Prove that $ED = EC$. It will then be easily shewn that $BA + AC$ is less than $BE + EC$.

19. Of equal triangles the equilateral has the least perimeter.

20. Of equal quadrilateral figures on the same base, that which has the least perimeter must have the angles not adjacent to the base equal to one another.

Let $ABCD$ be a quadrilateral figure on the base AD; and let the angle B be greater than the angle C. Produce AB, DC to meet in E. Take $EF = EB$ and let G be any point in CF.

Join GB, and make the angle $GBH = BGC$ and $BH = GC$.

Join GH: the triangles GBH, BGC are equal in every respect.

Also ABG is greater than ABF, and GBH than EBF.　　[I. 10. Cor. 2. I. 11.
Hence $ABG + GBH$ are greater than $2R$; and the pentagon $ABHGD$ has a re-entering angle at B. Also the area of this pentagon is equal to that of $ABCD$ and so also is its perimeter.

Join AH. The quadrilateral figure $AHGD$ is greater than $ABCD$ and has a less perimeter. Therefore $ABCD$ is not the figure of least perimeter.

21. Of equal quadrilateral figures the square has the least perimeter.

22. Of equal polygons of the same number of sides that which is regular has the least perimeter.

The perimeter of a square is less than that of the equilateral triangle of equal area.

Let ABC be an equilateral triangle. In AC take any point D.

Join DB. Through A draw AE parallel to DB. And let BED be the isosceles triangle described on BD.

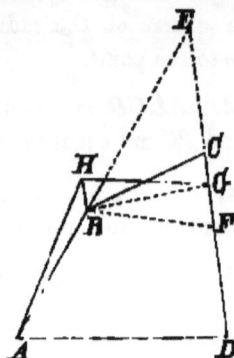

Then $BED = BAD$, and $BE + ED$ are less than $BA + AD$.

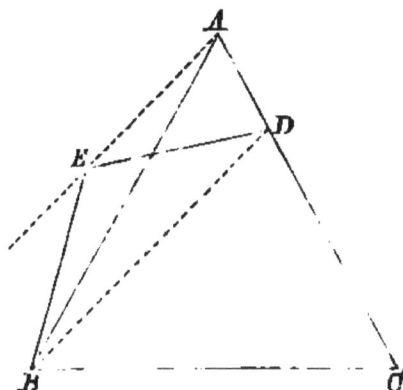

Hence $CBED$ is a quadrilateral figure equal to BAC in area but of less perimeter.

And the perimeter of the square equal to $CBED$ is less than that of $CBED$.

23. The perimeter of a regular pentagon is less than that of the square of equal area.

24. The perimeter of a regular polygon of $(n + 1)$ sides is less than that of a regular polygon of n sides of equal area.

25. The perimeter of a circle is less than that of any figure of equal area.

26. The cells of a honeycomb are hexagonal. Why should this be so?

27. What is the locus of the vertex of a triangle of constant area on a given base?

1. CONSTRUCT a parallelogram equal to a given triangle, and having an angle equal to a given angle.

Bisect the base of the triangle, and through the point of bisection draw a line making with the base an angle equal to the given angle.

Through the vertex draw a straight line parallel to the base. Two sides of the required parallelogram are now determined.

2. Construct on a given base a triangle equal to a given triangle.

Let ABC be the given triangle.

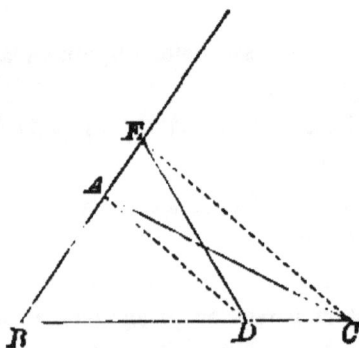

Take BD equal to the given base.

Join AD. Through C draw CE parallel to AD meeting BA produced in E.

Join DE. BDE is the triangle required.

For triangle ADE = triangle ADC,
and therefore the triangle BDE = the triangle BAC.

By the help of this proposition the area of a triangle may be found by construction alone.

For the measure of the triangle $BDE = \dfrac{BD \times \text{the altitude}}{2}$.

Hence if BD be taken equal to twice the unit of length.

The measure of the triangle $BDE = \dfrac{2 \times \text{altitude}}{2} = \text{altitude}$.

The triangle being of course expressed in superficial, and the altitude in linear units. The altitude being then measured the area of the triangle is found.

3. Find by above method the area of a triangle the sides of which are 3, 4, and 5 inches.

4. Find the area of the triangle having sides of 3 and 4 inches and enclosing an angle of 30 degrees.

5. Construct with a given altitude a triangle equal to a given triangle.

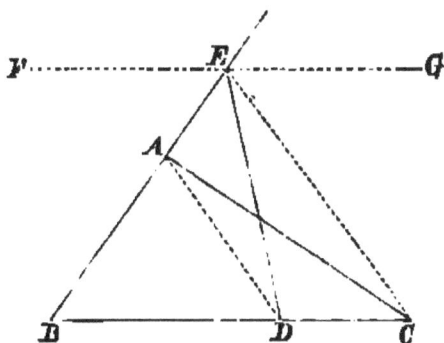

Draw FG parallel to BC at a perpendicular distance from it equal to the given altitude. Let it meet BA or BA produced in E. Join CE. Through A draw AD parallel to EC. Join ED. BDE is the triangle required.

If the altitude be taken equal to twice the unit of length the linear measure of the base will be equal to the superficial measure of the triangle.

6. Find the area of a triangle the sides of which are 3, 3¼, and 1¼ inches.

7. Find the area of a triangle having sides equal to 5 and 4 inches containing an angle of 150 degrees.

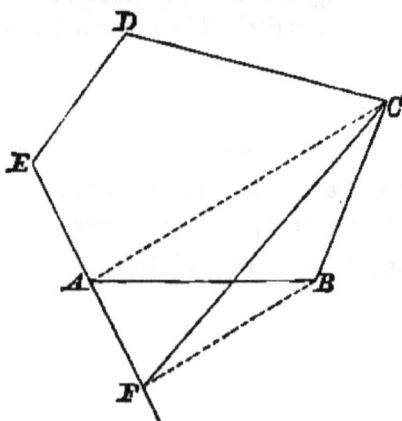

8. To construct a triangle equal to a given polygon.

Let *ABCDE* be a polygon of five sides.

Join *AC*. Through *B* draw *BF* parallel to *CA* meeting *EA* produced in *F*. Join *FC*.

Then the triangle *AFC* = triangle *ABC*.

And therefore the quadrilateral figure *FCDE* = the pentagon *ABCDE*.

The given construction enables us to transform any polygon into an equal polygon with one side less. Thus every polygon may, by repetition of the process, be reduced to an equivalent triangle.

Hence if required its area may be found by construction alone.

9. To construct a square equal to the sum of two given squares.

10. To construct a square equal to the difference of two given squares.

11. To divide a given line so that the sum of the squares of the parts may be equal to a given square. What are the necessary conditions?

12. To divide a given line so that the rectangle contained by the segments may be equal to a given square. The conditions?

13. Bisect a given triangle by a straight line drawn through a point in one of its sides.

BOOK IV.

THEOREM I.

PARALLEL straight lines which cut off equal parts from
one of two straight lines cut off
also equal parts from the other.

Let AB, AC be straight
lines meeting at A, AD, DE
equal parts cut off from AB; let
DF and EG be parallel straight
lines passing through D and E
and meeting AC in F and G.

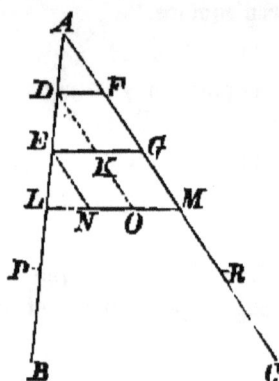

Then shall $AF = FG$.

Draw DK parallel to FG.

The triangles ADF, DEK
are equiangular.   [I. 8.

And	$AD = DE$.	
Therefore	$AF = DK$.	[I. 15.
But	$DK = FG$.	[I. 16.
Therefore	$AF = FG$.	

COR. 1. Similarly, if EL, LP, &c. be all equal to AD;
GM, MR, &c. shall be all equal to AF.

Cor. 2. From the congruent triangles ADF, DEK,

$$DF = EK.$$

Therefore $\quad EK = KG = DF.$

Also $\quad LN = NO = OM = DF.$

Cor. 3. Therefore as many parts as there are in AL each equal to AD, so many are there in AM each equal to AF, and so many in LM each equal to DF.

Theorem II.

If a straight line LM be drawn parallel to BC one of the sides of the triangle ABC, then shall

$$\frac{AL}{AB} = \frac{LM}{BC} = \frac{MA}{CA}.$$

Divide AL and AB into a number of equal parts as AD, and through the points of section draw straight lines parallel to BC or LM.

These shall divide AM, AC into a like number of equal parts, as AF. [IV. 1. Cor. 3.

Then if AL contain AD m times, so that $AL = m . AD$, then $\quad AM = m . AF$, and also $\quad LM = m . DF$.

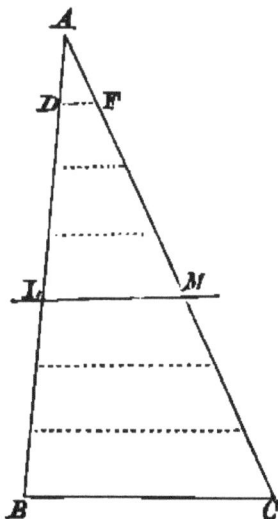

Again, if AB contain AD n times, so that $AB = n . AD$, then $\quad AC = n . AF$, and $BC = n . DF$.

6

Hence $\qquad \dfrac{AL}{AB} = \dfrac{m \cdot AD}{n \cdot AD} = \dfrac{m}{n}$.

Also $\qquad \dfrac{LM}{BC} = \dfrac{m \cdot DF}{n \cdot DF} = \dfrac{m}{n}$,

and $\qquad \dfrac{MA}{CA} = \dfrac{m \cdot AF}{n \cdot AF} = \dfrac{m}{n}$.

Therefore $\qquad \dfrac{AL}{AB} = \dfrac{LM}{BC} = \dfrac{MA}{CA}$.

Cor. Since $\dfrac{AL}{AB} = \dfrac{AM}{AC}$, therefore also

$$\dfrac{AL}{AB - AL} = \dfrac{AM}{AC - AM},$$

or $\qquad \dfrac{AL}{LB} = \dfrac{AM}{MC}$,

Conversely, if LM cut AB, AC so that $\dfrac{AL}{AB} = \dfrac{AM}{AC}$,

or so that $\qquad \dfrac{AL}{LB} = \dfrac{AM}{MC}$,

then LM is parallel to BC.

THEOREM III.

The sides about the angles of equiangular triangles are proportional.

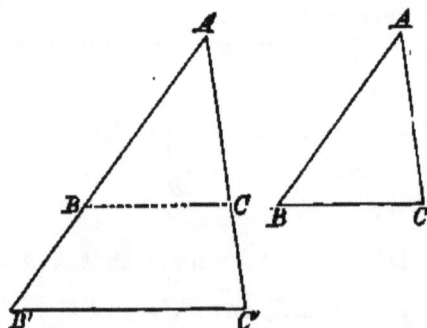

Let ABC, $A'B'C'$ be equiangular triangles.

Then shall $AB : A'B' = BC : B'C' = CA : C'A'$.

Apply the triangle ABC to the triangle $A'B'C'$, so that A may be on A', and AB on $A'B'$, AC must then fall on $A'C'$, since the angles A and A' are equal.

The points B and C will fall somewhere on the lines $A'B'$, $A'C'$.

And since $B = B'$

$\qquad BC$ must be parallel to $B'C'$. $\qquad$ [1. 7.

Therefore

$\qquad AB : A'B' = BC : B'C' = CA : C'A'$. $\qquad$ [IV. 2.

THEOREM IV.

If two triangles ABC, $A'B'C'$, have the angle A equal to the angle A', and the sides about the equal angles proportional, so that $AB : A'B' = AC : A'C'$, these triangles shall be equiangular.

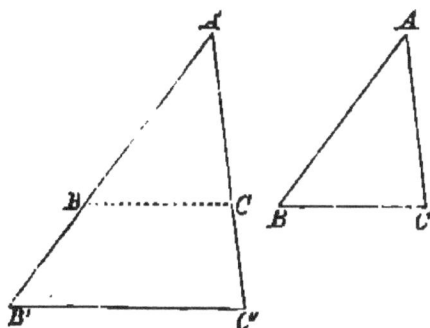

Apply the triangle ABC to the triangle $A'B'C'$, so that

the point A may be on A', and AB on $A'B'$, AC must then fall on $A'C'$ since the angles A and A' are equal.

And since $\qquad AB : A'B' = AC : A'C'$,

$\qquad\qquad BC$ is parallel to $B'C'$. $\qquad$ [IV. 2.

Therefore the remaining angles of the one triangle are equal to the remaining angles of the other each to each. [I. 8.

THEOREM V.

If in the triangles ABC, $A'B'C'$,

$$AB : A'B' = BC : B'C' = CA : C'A'$$

the triangles ABC, $A'B'C'$ are equiangular.

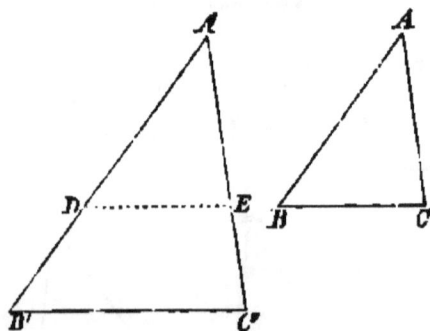

On $A'B'$ take $A'D = AB$, and through D draw DE parallel to $B'C'$. The triangle $A'DE$ is equiangular to the triangle $A'B'C'$. $\qquad$ [I. 8.

Therefore $\qquad A'D : A'B' = DE : B'C'$. $\qquad$ [IV. 3.

But $A'D$ being equal to AB, we have from hypothesis

$$A'D : A'B' = BC : B'C'.$$

Therefore $\quad DE : B'C' = BC : B'C'$.

Therefore $\quad\quad\quad DE = BC$.

Similarly $\quad\quad\quad EA' = CA$.

Therefore the triangles $A'DE$ and ABC are congruent and equiangular. [I. 13.

But $\quad A'DE$ is equiangular to $A'B'C'$;

Therefore ABC is equiangular to $A'B'C'$.

Relation of Areas of Figures.

THEOREM VI.

Triangles which have one angle of the one equal to one angle of the other, are to each other as the products of the sides containing the equal angle.

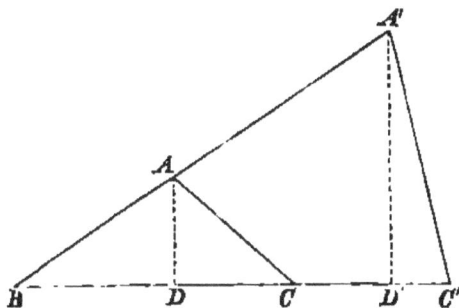

Let the triangles ABC, $A'BC'$ have equal angles at B.

Then shall $ABC : A'BC' = AB . BC : A'B . BC'$.

Place the triangles so that the sides containing the equal angle in the one may be upon those containing the equal angle in the other; and let fall the perpendiculars AD, $A'D'$.

Then the triangle $ABC = \dfrac{1}{2} AD . BC.$ [III. 7.

And the triangle $A'BC' = \dfrac{1}{2} A'D' . BC'.$ [III. 7.

Therefore

triangle ABC : triangle $A'BC' = AD . BC : A'D' . BC'.$

But $AD : A'D' = AB : A'B.$ [IV. 2.

Therefore

triangle ABC : triangle $A'BC' = AB . BC : A'B . BC'.$

Cor. 1. Parallelograms which have one angle of the one

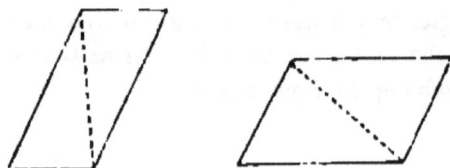

equal to one angle of the other are to one another as the products of the sides containing the equal angle.

Cor. 2. If the triangle $ABC =$ the triangle $A'BC'$, then

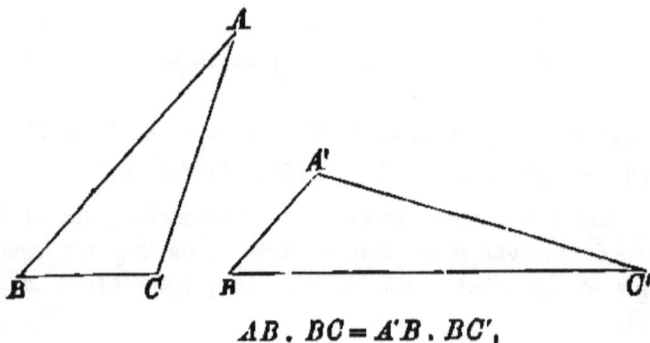

$$AB . BC = A'B . BC',$$

and therefore $AB : A'B = BC' : BC.$

i.e. the triangles ABC, $A'BC'$ have their sides about the equal angles reciprocally proportional.

Cor. 3.   If $AB : A'B = BC' : BC,$

$$AB . BC = A'B . BC',$$

and therefore the triangle $ABC =$ triangle $A'BC'$.

Or if ABC, $A'BC'$ have their sides about their equal angles reciprocally proportional, these triangles are equal.

DEF. Similar figures are those which have their several angles equal, each to each, and the sides about the equal angles proportionals.

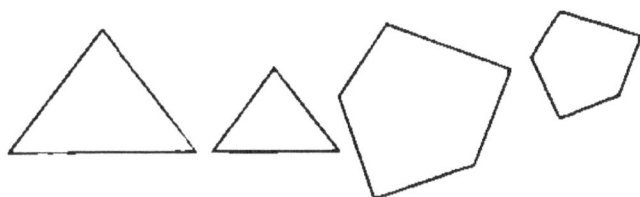

It has been shewn that equiangular triangles have their sides about their equal angles proportionals; triangles therefore which are equiangular are necessarily similar, but this is not the case with other figures. A square and an oblong are equiangular but not similar, and parallelograms equiangular but not similar are of common occurrence.

Theorem VII.

Similar triangles are to one another as the squares described upon their homologous sides; (or in the duplicate ratio of their homologous sides).

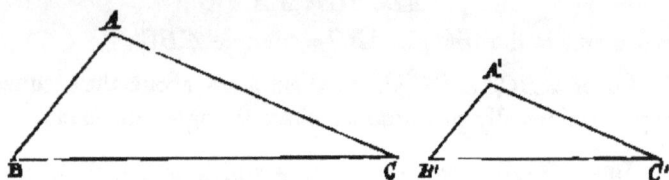

Let ABC, $A'B'C'$, be similar triangles, they shall be to one another as the squares on BC and $B'C'$.

For since ABC, $A'B'C'$ are similar triangles,

$$AB : BC = A'B' : B'C'.$$

Therefore $AB . BC : BC^2 = A'B' . B'C' : B'C'^2$;

or $\qquad AB . BC : A'B' . B'C' = BC^2 : B'C'^2.$

But since $B = B'$,

the triangle ABC : triangle $A'B'C' = AB . BC : A'B'. B'C'$.

[ɪᴠ. 6.

Therefore the

triangle ABC : triangle $A'B'C' = BC^2 : B'C'^2$.

Note. Duplicate ratio is thus defined:

When three magnitudes are proportionals, the first is said to have to the third the duplicate ratio of that which it has to the second.

Thus, if $A : B = B : C$, A has to C the duplicate ratio of that which it has to B. It will be seen that this ratio is equal to that of A^2 to B^2.

THEOREM VIII.

Similar parallelograms are to one another as the squares upon their homologous sides; (or in the duplicate ratio of their homologous sides).

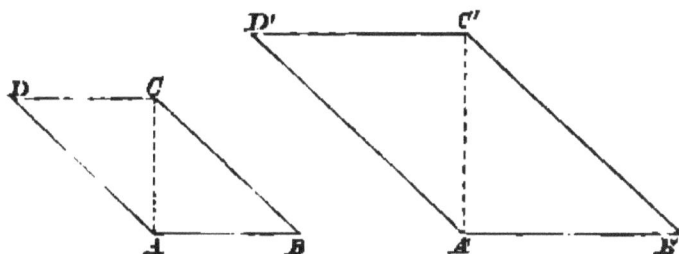

For they can be divided into similar triangles.

THEOREM IX.

Similar polygons are to one another in the duplicate ratio of their homologous sides.

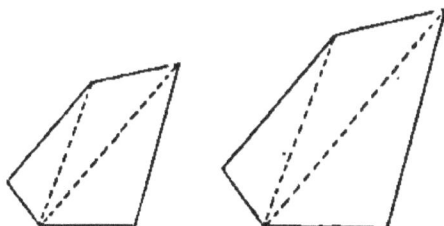

For they may always be divided into a like number of similar triangles, each triangle in the first bearing a constant ratio to the corresponding triangle in the second.

The proof we leave as an exercise for the learner.

The proposition that "Similar figures are to one another
in the duplicate ratio of their homologous sides" is true of
curvilinear figures as well as of rectilinear. Thus two circles
are to one another as the squares on their diameters, or on their
radii. The proof of this is too difficult to be given here.

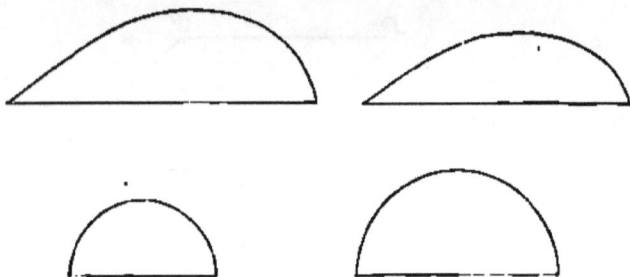

Properties of a Triangle.

Theorem X.

In the triangle *ABC*, if *AD* be drawn from *A* to *D* the
middle point of *BC*, and *BE*, *CF* be similarly drawn from
B and *C* to the middle points of the opposite sides; *AD*, *BE*,
CF shall intersect in one point.

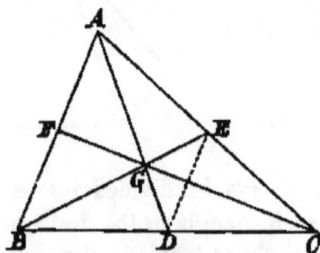

Let *AD* and *BE* intersect in *G*. Join *DE*.

Then since $BD : DC = AE : EC,$

DE is parallel to AB.   [IV. 2.

Hence the triangles AGB, DGE are equiangular and similar.

Therefore $AG : GD = AB : DE,$ [IV. 3.

But $AB : DE = BC : DC,$

whence $AB = $ twice $DE.$

Therefore $AG = $ twice $GD.$

That is, the point of intersection of AD and BE is so situated on AD, that its distance from A is twice as great as its distance from D.

Similarly the point of intersection of AD, CF is so situated on AD, that its distance from A is twice as great as its distance from D.

Therefore these points coincide.

This is a proposition of great importance in Mechanics. The point G is the centre of inertia of the triangle ABC.

THEOREM XI.

In the triangle ABC, if AD be drawn from A perpendicular to BC, and BE, CF be similarly drawn from B and C perpendicular to the opposite sides; AD, BE, CF shall intersect in one point.

Through A, B, and C draw GH, HK, KG parallel to BC, CA, AB respectively.

The figures AK, BG, CH are parallelograms.

Therefore $AH = BC = AG$,   [I. 16.

or A bisects GH.

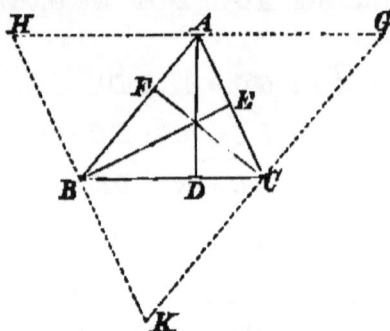

Similarly B and C bisect HK, KG.

Therefore every point on AD is equidistant from G and H,   [I. 19.

and every point on BE is equidistant from H and K.   [I. 19.

Therefore the point of intersection of AD, BE is equidistant from G and K.

Therefore it is a point on CF.   [I. 21.

Theorem XII.

If ABC be a right-angled triangle having a right angle at A, and from A, AD be let fall perpendicular to BC, the triangles ABD, ACD shall be similar to the whole triangle and to one another.

For in the triangles ABD, ABC the right angle ADB is equal to the right angle BAC, and the angle at B is common; therefore the triangles are equiangular and similar.

In the same way ABC, ACD are equiangular and similar.

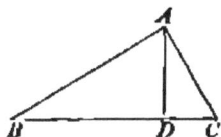

Therefore also ABD, ACD are equiangular and similar.

Cor. 1. In the similar triangles BDA, ADC,

$$BD : DA = AD : DC;$$

therefore $BD \cdot DC = AD^2$.

Cor. 2. In the similar triangles BCA, ACD,

$$BC : CA = AC : CD;$$

therefore $BC : CD = BC^2 : CA^2$.

Similarly $BC : BD = BC^2 : AB^2$;

therefore $BC : CD + BD = BC^2 : CA^2 + AB^2$,

but $BC = CD + BD;$

therefore $BC^2 = CA^2 + AB^2$,

or, the square on the hypothenuse is equal to the squares on the other two sides.

Cor. 3. Any figure on BC is equal to the similar and similarly described figures on BA and AC.

For these figures are to one another as the squares on BC, BA, and AC.   [IV. 9.

Theorem XIII.

If on each of the sides of the right-angled triangle ABC, as diameter, a semicircle be described, the crescents $ADBE$, $AFCG$ are together equal to the triangle ABC.

For the semicircles CFA, ADB, and BAC are similar

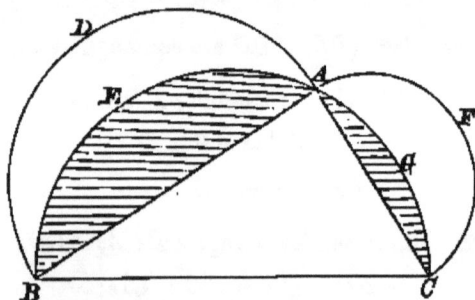

figures described on the sides and the hypothenuse of the right-angled triangle BAC:

Therefore the semicircles CFA and ADB are equal to the semicircle BAC. [IV. 9.

Take from each the segments BEA, AGC.

Therefore the crescents $ADBE$, $AFCG$ are equal to the triangle ABC.

This theorem possesses great historical interest, as the first example of the quadrature of a curvilinear figure. It is due to Hippocrates, 450 B.C.

Theorem XIV.

If ABC be a triangle, and from B and C BE, CF be let fall perpendicular to CA, AB, or these produced,

Then shall $BA \cdot AF = CA \cdot AE.$

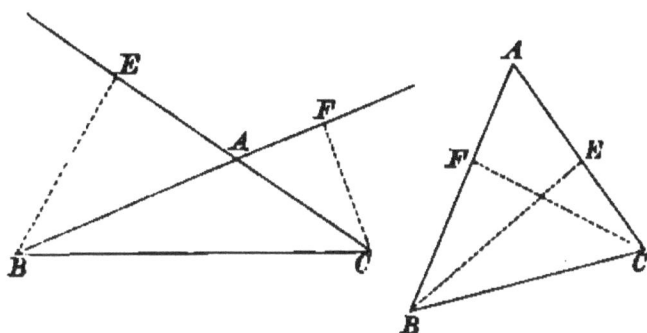

For the right-angled triangles ABE, ACF have a common or equal angle at A, and are therefore equiangular and similar.

Therefore $\quad BA : AE = CA : AF.$

Therefore $\quad BA \cdot AF = CA \cdot AE.$

Cor. Similarly

$$CB \cdot BD = AB \cdot BF.$$

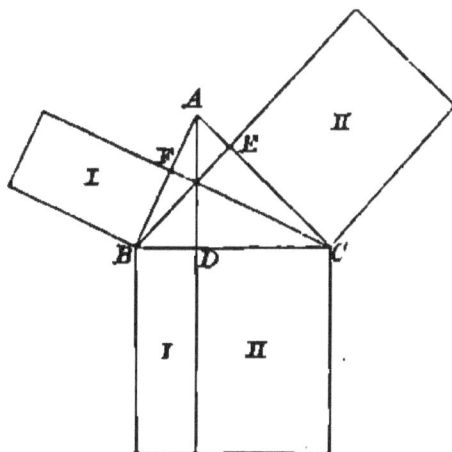

And $BC \cdot CD = AC \cdot CE.$

Therefore $BC(BD + CD) = AB \cdot BF + AC \cdot CE,$

or $BC^2 = AB \cdot BF + AC \cdot CE.$

THEOREM XV.

If A be a right angle

$$BC^2 = CA^2 + AB^2.$$

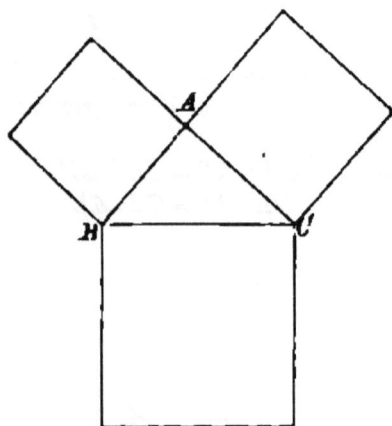

For it has been shewn that

$$DC^2 = AC \cdot CE + AB \cdot BF,$$

where E and F are the feet of the perpendiculars from B and C on CA and AB.

But if A be a right angle E and F both coincide with A, and $CE = CA,$ and $BF = BA.$

Therefore $BC^2 = CA^2 + AB^2.$

Theorem XVI.

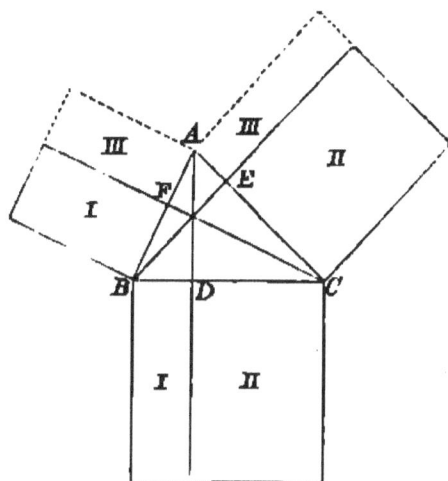

If A be less than a right angle

$$BC^2 = CA^2 + AB^2 - 2CA \cdot AE.$$

For, as before,

$$BC^2 = AD \cdot BF + AC \cdot CE.$$

But $\quad AB \cdot BF = AB^2 - BA \cdot AF,$

and $\quad\quad AC \cdot CE = CA^2 - CA \cdot AE.$

Also $\quad\quad BA \cdot AF = CA \cdot AE.$

Therefore $BC^2 = CA^2 + AB^2 - 2CA \cdot AE.$

Theorem XVII.

If A be greater than a right angle

$$BC^2 = CA^2 + AB^2 + 2CA \cdot AE.$$

7

For as before

$$BC^2 = AB \cdot BF + AC \cdot CE,$$

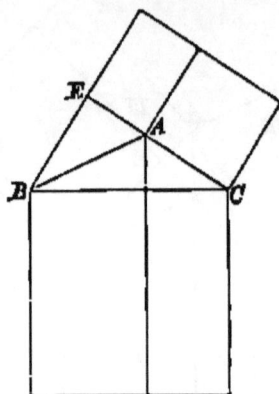

but $\qquad AB \cdot BF = AB^2 + BA \cdot AF,$

and $\qquad AC \cdot CE = CA^2 + CA \cdot AE,$

and also $\qquad BA \cdot AF = CA \cdot AE.$

Therefore $\quad BC^2 = CA^2 + AB^2 + 2CA \cdot AE.$

THEOREM XVIII.

If two chords of a circle intersect, the rectangles contained by their segments are equal.

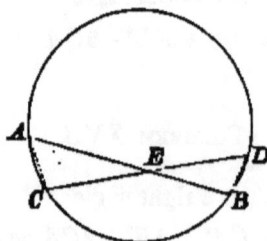

N.B. A segment of a line is a part cut off from a line; the segments spoken of above are to be reckoned from the circumference to the point of intersection and again from the point of intersection to the circumference.

Let AB, CD intersect in E.

Then shall $AE.EB = CE.ED$.

First, let E lie within the circumference of the circle.

Join AC, BD.

Then $CAB = CDB$,

being on the same arc CB,   [II. 14.

and $AEC = DEB$.   [I. 2.

Therefore the triangles AEC, DEB are equiangular and similar.

Therefore $AE : EC = DE : EB$,   [IV. 3.

whence $AE.EB = CE.ED$.

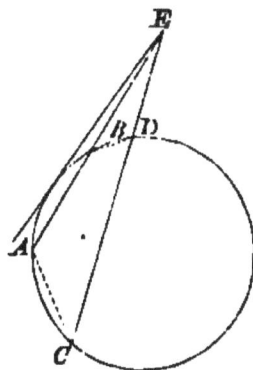

Secondly, let E be without the circumference of the circle.

Join AC, BD.

Then $EBD = ACE$, since each is supplemen-   [II. 15, Cor. 2.
tary to ABD, and E is common to the triangles AEC, DEB.

Therefore these triangles are equiangular and similar.

Therefore $AE : EC = DE : EB$,   [IV. 3.

whence $AE . EB = CE . ED$.

Cor. Let AE be turned about E so as to become more
remote from EC. The points A and B will continually
approach one another and at last coincide. We shall then
have

$$AE = EB \text{ and } AE . EB = AE',$$

and AE will be the tangent at A.

And

$$CE . ED = AE',$$

or the rectangle contained by the segments of a chord drawn
through an external point is equal to the square on the tangent drawn from that point.

We append a proof independent of the method of limits.

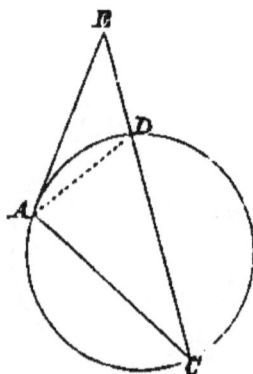

Let CDE be a chord passing through E, EA the tangent
drawn from E. Join AD, AC.

In the triangles CEA, AED, the angle at E is common; and the angle $DAE = ACE$, since EA touches the circle. [II. 16.

Therefore the triangles CEA, AED are equiangular and similar.

Whence $\qquad CE : EA = AE : ED,$ [IV. 3.

and therefore $\qquad CE . ED = AE^2.$

THEOREM XIX.

If one angle of a triangle be bisected by a straight line which cuts the base, the sides of the triangle shall be to one another as the segments of the base.

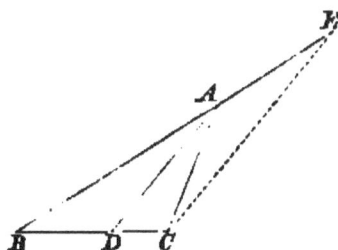

In the triangle ABC, let A be bisected by AD.

Then $\qquad BA : AC = BD : DC.$

Produce BA, and through C draw CE parallel to DA meeting BA produced in E.

Then the exterior angle $BAD = AEC$, [I. 8.

and the alternate angle $DAC = ACE$, [I. 8.

but $\qquad BAD = DAC.$

Therefore $\qquad AEC = ACE,$

whence $\qquad AC = AE.$

And since AD is parallel to CE,

$$BA : AE = BD : DC.$$ [IV. 2.

Therefore $BA : AC = BD : DC.$

Conversely, if $BA : AC = BD : DC$, DA bisects BAC.

The same construction being made

$$BA : AE = BD : DC,$$ [IV. 2.

but by hypothesis

$$BA : AC = BD : DC.$$

Therefore $AE = AC,$

whence $BAD = CAD.$

THEOREM XX.

If the exterior angle of a triangle be bisected by a straight line which cuts the base, the sides of the triangle shall be to one another as the segments of the base.

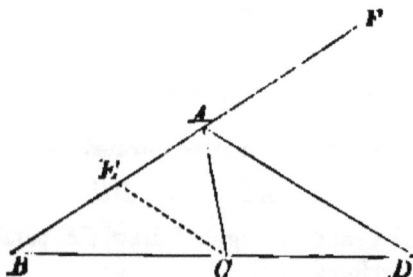

In the triangle ABC, let FAC be bisected by AD.

Then shall $BA : AC = BD : DC.$

Through C draw CE parallel to DA.

Then as in the preceding proposition $AE = AC$.

And $BA : AE = BD : DC.$ [IV. 2.

Therefore $BA : AC = BD : DC.$

Conversely, if $BA : AC = BD : DC$, DA bisects FAC.

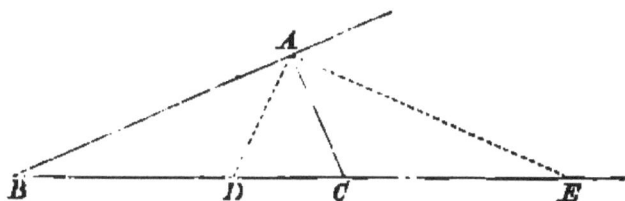

Cor. If AD, AE bisect the interior and exterior angles at A,

$$BA : AC = BD : DC,$$

also $$BA : AC = BE : EC.$$

Therefore $BD : DC = BE : EC$,

or $$BE : EC = DE - ED : ED - EC.$$

That is EB, ED, EC are in Harmonic Progression. .

Similarly BE, BC, BD are in Harmonic Progression.

Harmonic Section.

THEOREM XXI.

If a given straight line BC be divided in any ratio not equal to unity in the point D, another point E may be found in BC produced such that $BE : EC = BD : DC.$

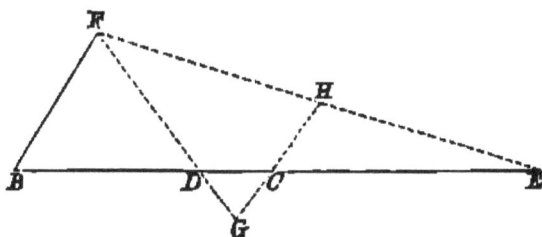

Draw any line BF. Join FD and produce it, through C draw GCH parallel to BF, meeting FD in G.

Make $CH = CG$. Join FH. Then since BF is not equal to CH, BC, FH being produced must meet in some point E.

Then $\qquad BE : EC = FB : HC \qquad\qquad$ [IV. 2.

$$= FB : GC$$

$$= BD : DC. \qquad\qquad \text{[IV. 2.}$$

The line BC is said to be harmonically divided.

THEOREM XXII.

If M be the middle point of a line AB which is divided harmonically in C and D, then shall the square on AM be equal to the rectangle $MC \cdot MD$.

—————————————————————————
$A \qquad\qquad M \qquad C \qquad\quad B \qquad\qquad\qquad D$

For $\qquad\qquad \dfrac{AC}{CB} = \dfrac{AD}{DB}$,

or $\qquad\qquad \dfrac{AM + MC}{AM - MC} = \dfrac{MD + AM}{MD - AM}$.

Therefore, $\qquad \dfrac{AM}{MC} = \dfrac{MD}{AM}$,

or $\qquad\qquad AM^2 = MC \cdot MD$.

COR. $\qquad DA, DM, DC, DB$ are proportionals.

For $\qquad DA \times DB = (DM + MA)(DM - MA)$

$$= DM^2 - MA^2$$

$$= DM^2 - MC \cdot MD$$

$$= DM(DM - MC)$$

$$= DM \cdot DC;$$

$\therefore DA : DM = DC : DB$.

THEOREMS.

1. AB, AC, AD, are lines drawn through A, EFG, KHL are parallel lines meeting them. Shew that
$$EF : FG = KH : HL,$$
and find all the proportions which the lines of the figure afford.

2. Any three lines are cut by three parallel lines, shew that they are divided proportionally.

3. AB, AC are drawn through A; from B draw BC to any point C on AC, and from C draw CD to any point D on AB. Draw DE parallel to BC, and EF parallel to CD. Shew that AD is a mean proportional between AB and AF.

4. The distance of a point P from a given line AB is always in a constant ratio to its distance from another line AC; find the locus of P.

5. From points on the side of an equilateral triangle at distances 2, 4, 8 from one of the base angles perpendiculars are let fall on the base. Find the lengths which they intercept.

6. ABC, $A'B'C'$ are triangles having equal angles at B, and at the angles C and C' supplementary. Then
$$BA : AC = BA' : A'C'.$$

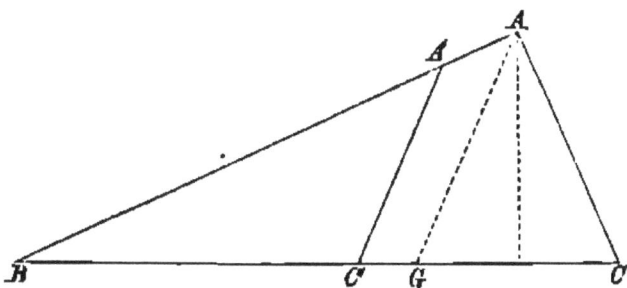

Place the triangles so that their equal angles coincide, and through A draw AG parallel to $A'C'$. Then
$$AGC = A'C'C = ACG, \text{ and } AG = AC.$$

But $\qquad BA : AG = BA' : A'C'.$ $\qquad$ [IV. 2.

Therefore $\qquad BA : AC = BA' : A'C'.$

7. If two triangles have one angle equal, and the sides about a second angle proportional, their remaining angles are either equal or supplementary.

Make the same construction as in 6. Shew that $AC = AG$. Then AC either coincides with AG, or is equally remote from the perpendicular on the other side. Compare I. Ambiguous case, and Constructions I. 7, 2.

Hence if ACB, $A'C'B$ be either both acute or both obtuse, or if one of them be a right angle, the triangles are similar.

8. ABC is a triangle, a line is drawn meeting BC, CA, AB, or those produced in D, E, and F. Shew that

$$AF . BD . CE = FB . DC . EA.$$

9. ABC is an equilateral triangle, P a point lying between B and C on the circumference of the circumscribed circle.

Shew that $PA = PB + PC$. $\qquad$ [II. 10, IV. 3.

10. In Construction 8, shew that the triangles ABD, ACD, CBD are proportional: also that

$$ABD : CBD = \text{sq. on } AB : \text{sq. on } AC.$$

11. The semicircle described upon the perpendicular of an equilateral triangle is to that on the side as 3 to 4.

12. O is the centre of a circle, AO its radius. On AO as diameter another circle is described, and any common chord is drawn through A. Shew that the segments which it cuts off from the two circles are in the ratio of 4 to 1.

It may be assumed that segments which contain equal angles are similar figures.

13. *AB* is the diameter of a circle, *CD* a perpendicular upon it from any point in the circumference. The semicircle on *AB* is equal to the semicircles on *AC* and *CB* together with the circle on *CD*.

14. *ABC* is a triangle, *D, E, F,* the middle points of *BC, CA, AB*.

  1. If $AD = BE$ then $A = B$.

  2. If *GH* be any line parallel to *BC* the locus of its middle point is *AD*.

  3. The locus of the intersection of *BH* and *GC* is also *AD*.

  4. The triangle constructed with the sides *AD, BE, CF* is to the original triangle as 3 to 4.

  5. The sum of the squares on *AD, BE, CF* is to the sum of the squares on the sides of the triangle as 3 to 4.

15. A triangle may be divided into three equal parts by lines drawn from a point within it to the angles.

16. *ABC* is an equilateral triangle, *AD* the perpendicular from *A*, *DG* the perpendicular from *D* on *AB*. Shew that *GB* is one-fourth of *AB*. Determine the ratio of the squares on *AG* and *AD*.

17. *ABC* is a triangle, *A* a right angle, *AD* the perpendicular from *A* on *BC*. Shew that

$$BC^2 : BA^2 : AC^2 \text{ as } BC : BD : DC.$$

18. *ABC* is any triangle, *AD* the perpendicular on *BC*; shew that

$$AB^2 - AC^2 = DB^2 - DC^2.$$

19. Find the area of the triangle whose sides are 17, 15, and 8.

20. Shew that IV. 14 may be deduced from IV. 18.

21. ABC is a triangle, AD the perpendicular on BC, AOE the diameter of the circle about ABC. Shew that
$$BA \cdot AC = EA \cdot AD.$$

22. $ABCD$ is a quadrilateral figure in a circle. Shew that
$$AC \cdot BD = AB \cdot CD + BC \cdot DA.$$

Make the angle $BAE = CAD$. The triangles ABE, ACD are similar, as also the triangles ADE, ACB.

Therefore
$$AB : BE = AC : CD \text{ and } AB \cdot CD = BE \cdot AC,$$
also $AD : DE = AC : CB$ and $AD \cdot CB = DE \cdot AC$.

23. ABC is a triangle, AD bisects A and meets the base in D and the circumscribed circle in E; BAD, EAC are similar triangles, and
$$BA \cdot AC = BD \cdot DC + AD^2.$$

24. What result is obtained when AD bisects the exterior angle at A?

25. In the triangle ABC, $AC : BC = 2 : 1$. CD, CE bisect the interior and exterior angles at C, the triangles CBD, ACD, ABC, CDE are as $1 : 2 : 3 : 4$.

26. A and B are two fixed points, P any point. Find the locus of P, when

1. $PA : PB$ is a constant ratio.
2. $PA^2 + PB^2 = $ a constant.
3. $PA^2 - PB^2 = $ a constant.

27. Find the locus of a point from which tangents drawn to two fixed circles are equal.

28. Find the locus of a point at which two given circles subtend equal angles.

29. Find a point at which three given circles subtend equal angles.

CONSTRUCTIONS.

1. To trisect a given line.

2. To divide a given line into 5 parts, or any number of parts.

3. To find a fourth proportional to three given straight lines.

4. To find a third proportional to two given straight lines.

5. To find a mean proportional to two given straight lines.

Place them in one straight line, on this describe a semi-circle, and at the point where the lines join erect a perpendicular to meet the circumference. This will be the mean proportional required. [IV. 12.

Construct a square equal to a given rectangle.

6. To divide a line into three parts which are to one another as 3 : 5 : 7 or as $m : n : p$.

7. To divide a line AB in extreme and mean ratio, i. e. so that $AB : AC = AC : CB$ or $AB . BC = AC^2$.

On AB describe a square $ABDE$, bisect EA in F. Join FB, and produce FA to G, making $FG = FB$. On AG describe a square : the angle of this square determines the required point on AB.

The proof from III. 11 and III. 12.

8. To construct an isosceles triangle having each of the angles at the base double of the third angle.

Take any straight line AB, and divide it so that
$$AB . BC = AC^2.$$

With centre A and distance AB describe a circle, in which place $BD = AC$. Join AD; ABD is the triangle

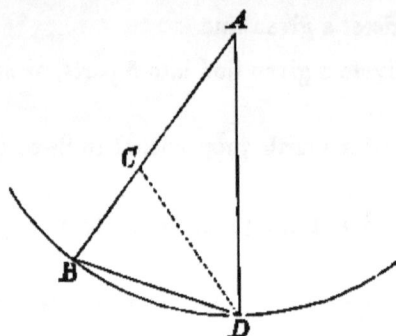

required. Join DC. ABD, DBC are similar triangles IV. 4. And therefore $DC = DB = AC$, whence $CAD = CDA$, and $ABD = DCB = $ twice CAD.

9. To inscribe a regular pentagon in a given circle.

The inscription of a regular figure in a circle depends upon our being able to divide the circumference into as many equal parts as the figure has sides.

We have seen how to divide a circumference into 3 parts or 4 parts, and since every arc can be bisected, we can divide a circumference into 6, 12, 24, &c. or 8, 16, 32, &c. parts.

The preceding construction gives us an angle CDB equal to one-fifth of two right angles. If therefore DC be produced to meet the circumference, it will intercept one-fifth part of it. Complete the construction.

10. To inscribe a regular decagon in a given circle.

11. To inscribe a regular quindecagon in a given circle.

12. On a given straight line to construct a polygon similar to a given polygon.

13. To inscribe a square in a triangle.

14. To inscribe a square in a given segment of a circle.

15. To describe a circle touching two straight lines and passing through a fixed point.

16. To describe a circle touching a straight line and passing through two fixed points.

17. To divide a triangle into two parts in the ratio of two given lines by a line drawn parallel to its base.

18. To produce a line AB to C so that AB, AC may be equal to a given square.

19. To construct a rectangle equal to a given square, and such that the difference of its sides may be equal to a given straight line.

20. To construct a figure similar to either of two given similar figures, and equal to their sum or their difference.

21. To construct a figure equal to one and similar to another given figure.

---

1. O is a point within the angle ACB. To draw through O a straight line AOB, so that it may be bisected in O.

2. A and B are two points on the same side of the straight line CD. Find in CD a point P such that $PA + PB$ may be the least possible.

3. In the figure of III. 12 shew that AL, CF, BK pass through one point.

4. An equilateral triangle is inscribed in a circle, if the intercepted arcs be bisected and the points of bisection joined, the sides of the triangles are trisected.

5. Through four given points to draw lines forming a square.

6. Shew that the circumference of a circle is more than three times and less than four times as great as the diameter.

7. Draw from the vertex of a triangle a line to the base such that it may be a mean proportional between the segments of the base.

8. AB is the diameter of a circle, DE a chord, OP a perpendicular on AB from any point O of DE. Shew that
$$AP.PB = DO.OE + OP^2.$$

9. The part of a variable tangent intercepted between two fixed tangents subtends a constant angle at the centre.

10. ABC is a triangle. AD, BE, CF are lines drawn through one point O to meet the opposite sides. Shew that
$$AF.BD.CE = FB.DC.EA.$$

11. The common tangents to two circles meet in a point O; and through O a chord $OKGMN$ is drawn meeting the circumferences in K, G, M and N. Shew that
$$OK.ON = OG.OM.$$

12. Given the perpendiculars to construct the triangle.

November, 1868.

LIST OF EDUCATIONAL BOOKS

PUBLISHED BY

MACMILLAN AND CO.,

16, BEDFORD STREET, COVENT GARDEN,

London, w.c.

CONTENTS.

MESSRS. MACMILLAN & CO. *beg to call attention to the accompanying Catalogue of their* EDUCATIONAL WORKS, *the writers of which are mostly scholars of eminence in the Universities, as well as of large experience in teaching.*

Many of the works have already attained a wide circulation in England and in the Colonies, and are acknowledged to be among the very best Educational Books on their respective subjects.

The books can generally be procured by ordering them through local booksellers in town or country, but if at any time difficulty should arise, MESSRS. MACMILLAN *will feel much obliged by direct communication with themselves on the subject.*

Notices of errors or defects in any of these works will be gratefully received and acknowledged.

LIST OF EDUCATIONAL BOOKS.

CLASSICAL.

ÆSCHYLI EUMENIDES. The Greek Text, with English Notes, and English Verse Translation and an Introduction. By BERNARD DRAKE, M.A., late Fellow of King's College, Cambridge. 8vo. 7s. 6d.

The Greek Text adopted in this Edition is based upon that of Wellauer, which may be said in general terms to represent that of the best manuscripts. But in correcting the Text, and in the Notes, advantage has been taken of the suggestions of Hermann, Paley, Linwood, and other commentators.

ARISTOTLE ON FALLACIES; OR, THE SOPHISTICI ELENCHI. With a Translation and Notes by EDWARD POSTE, M.A., Fellow of Oriel College, Oxford. 8vo. 8s. 6d.

Besides the doctrine of Fallacies, Aristotle offers either in this treatise, or in other passages quoted in the commentary, various glances over the world of science and opinion, various suggestions on problems which are still agitated, and a vivid picture of the ancient system of dialectics, which it is hoped may be found both interesting and instructive.

" It is not only scholarlike and careful; it is also perspicuous."—*Guardian.*

ARISTOTLE.—AN INTRODUCTION TO ARISTOTLE'S RHETORIC. With Analysis, Notes, and Appendices. By E. M. COPE, Senior Fellow and Tutor of Trinity College, Cambridge. 8vo. 14s.

This work is introductory to an edition of the Greek Text of Aristotle's Rhetoric, which is in course of preparation.

" Mr. Cope has given a very useful appendage to the promised Greek Text; but also a work of so much independent use that he is quite justified in his separate publication. All who have the Greek Text will find themselves supplied with a comment; and those who have not will find an analysis of the work."—*Athenæum.*

CATULLI VERONENSIS LIBER, edited by R. ELLIS, Fellow of Trinity College, Oxford. 18mo. 3s. 6d.

> " It is little to say that no edition of Catullus at once so scholarlike has ever appeared in England."—*Athenæum.*

> " Rarely have we read a classic author with so reliable, acute, and safe a guide."—*Saturday Review.*

CICERO.—THE SECOND PHILIPPIC ORATION. With an Introduction and Notes, translated from the German of KARL HALM. Edited, with Corrections and Additions, by JOHN F. B. MAYOR, M.A., Fellow and Classical Lecturer of St. John's College, Cambridge. Third Edition, revised. Fcap. 8vo. 5s.

> " A very valuable edition, from which the student may gather much both in the way of information directly communicated, and directions to other sources of knowledge."—*Athenæum.*

DEMOSTHENES ON THE CROWN. The Greek Text with English Notes. By B. DRAKE, M.A., late Fellow of King's College, Cambridge. Third Edition, to which is prefixed ÆSCHINES AGAINST CTESIPHON, with English Notes. Fcap. 8vo. 5s.

> The terseness and felicity of Mr. Drake's translations constitute perhaps the chief value of his edition, and the historical and archæological details necessary to understanding the *De Coronâ* have in some measure been anticipated in the notes on the Oration of Æschines. In both, the text adopted in the Zurich edition of 1851, and taken from the Parisian MS., has been adhered to without any variation. Where the readings of Bekker, Dissen, and others appear preferable, they are subjoined in the notes.

HODGSON.—MYTHOLOGY FOR LATIN VERSIFICATION. A Brief Sketch of the Fables of the Ancients, prepared to be rendered into Latin Verse for Schools. By F. HODGSON, B.D., late Provost of Eton. New Edition, revised by F. C. HODGSON, M.A. 18mo. 3s.

> Intending the little book to be entirely elementary, the Author has made it as easy as he could, without too largely superseding the use of the Dictionary and Gradus. By the facilities here afforded, it will be possible, in many cases, for a boy to get rapidly through these preparatory exercises : and thus, having mastered the first difficulties, he may advance with better hopes of improvement to subjects of higher character, and verses of more difficult composition.

JUVENAL, FOR SCHOOLS. With English Notes. By J. E. B.
MAYOR, M.A. New and Cheaper Edition. Crown 8vo.
[In the Press.

"A School edition of Juvenal, which, for really ripe scholarship, extensive
acquaintance with Latin literature, and familiar knowledge of Continental
criticism, ancient and modern, is unsurpassed, we do not say among English School-books, but among English editions generally."—*Edinburgh
Review.*

LYTTELTON.—THE COMUS of MILTON rendered into Greek
Verse. By LORD LYTTELTON. Extra fcap. 8vo. Second Edition.
5*s.*

— THE SAMSON AGONISTES of MILTON rendered into
Greek Verse. By LORD LYTTELTON. Extra fcap. 8vo. 6*s.* 6*d.*

MARSHALL.—A TABLE OF IRREGULAR GREEK VERBS,
Classified according to the Arrangement of Curtius's Greek
Grammar. By J. M. MARSHALL, M.A., Fellow and late Lec-
turer of Brasenose College, Oxford; one of the Masters in Clifton
College. 8vo. cloth. 1*s.*

MAYOR.—FIRST GREEK READER. Edited after KARL HALM,
with Corrections and large Additions by JOHN E. B. MAYOR, M.A.,
Fellow and Classical Lecturer of St. John's College, Cambridge.
Fcap. 8vo. 6*s.*

MERIVALE.—KEATS' HYPERION rendered into Latin Verse.
By C. MERIVALE, B.D. Second Edition. Extra fcap. 8vo.
3*s.* 6*d.*

PLATO.—THE REPUBLIC OF PLATO. Translated into English, with an Analysis and Notes, by J. LL. DAVIES, M.A., and
D. J. VAUGHAN, M.A. Third Edition, with Vignette Portraits
of Plato and Socrates, engraved by JEENS from an Antique Gem.
18mo. 4*s.* 6*d.*

ROBY.—A LATIN GRAMMAR for the Higher Classes in Grammar
Schools. By H. J. ROBY, M.A.; based on the "Elementary
Latin Grammar." *[In the Press.*

SALLUST.—CAII SALLUSTII CRISPI Catilina et Jugurtha. For use in Schools (with copious Notes). By C. MERIVALE, B.D. (In the present Edition the Notes have been carefully revised, and a few remarks and explanations added.) Second Edition. Fcap. 8vo. 4s. 6d.

The Jugurtha and the Catilina may be had separately, price 2s. 6d. each.

TACITUS.—THE HISTORY OF TACITUS translated into ENGLISH. By A. J. CHURCH, M.A., and W. J. BRODRIBB, M.A. With Notes and a Map. 8vo. 10s. 6d.

> The translators have endeavoured to adhere as closely to the original as was thought consistent with a proper observance of English idiom. At the same time it has been their aim to reproduce the precise expressions of the author. The campaign of Civilis is elucidated in a note of some length which is illustrated by a map, containing only the names of places and of tribes occurring in the work.

— THE AGRICOLA and GERMANY. By the same translators. With Maps and Notes. Extra fcap. 8vo. 2s. 6d.

THRING.—Works by **Edward Thring, M.A.**, Head Master of Uppingham School :—

— A CONSTRUING BOOK. Fcap. 8vo. 2s. 6d.

> This Construing Book is drawn up on the same sort of graduated scale as the Author's *English Grammar.* Passages out of the best Latin Poets are gradually built up into their perfect shape. The few words altered, or inserted as the passages go on, are printed in Italics. It is hoped by this plan that the learner, whilst acquiring the rudiments of language, may store his mind with good poetry and a good vocabulary.

— A LATIN GRADUAL. A First Latin Construing Book for Beginners. Fcap. 8vo. 2s. 6d.

> The main plan of this little work has been well tested.
> The intention is to supply by easy steps a knowledge of Grammar, combined with a good vocabulary: in a word, a book which will not require to be forgotten again as the learner advances.
> A short practical manual of common Mood constructions, with their English equivalents, form the second part.

— A MANUAL of MOOD CONSTRUCTIONS. Extra fcap. 8vo. 1s. 6d.

THUCYDIDES.—THE SICILIAN EXPEDITION. Being Books VI. and VII. of Thucydides, with Notes. A New Edition, revised and enlarged, with a Map. By the Rev. PERCIVAL FROST, M.A., late Fellow of St. John's College, Cambridge. Fcap. 8vo. 5s.

> This edition is mainly a grammatical one. Attention is called to the force of compound verbs, and the exact meaning of the various tenses employed.

WRIGHT.—Works by **J. Wright, M.A.**, late Head Master of Sutton Coldfield School :—

— HELLENICA; Or, a HISTORY of GREECE in GREEK, as related by Diodorus and Thucydides, being a First Greek Reading Book, with Explanatory Notes Critical and Historical. Second Edition, with a Vocabulary. 12mo. 3s. 6d.

> In the last twenty chapters of this volume, Thucydides sketches the rise and progress of the Athenian Empire in so clear a style and in such simple language, that the author doubts whether any easier or more instructive passages can be selected for the use of the pupil who is commencing Greek.

— A HELP TO LATIN GRAMMAR; Or, the Form and Use of Words in Latin, with Progressive Exercises. Crown 8vo. 4s. 6d.

> "Never was there a better aid offered alike to teacher and scholar in that arduous pass. The style is at once familiar and strikingly simple and lucid; and the explanations precisely hit the difficulties, and thoroughly explain them."—*English Journal of Education.*

— THE SEVEN KINGS OF ROME. An Easy Narrative, abridged from the First Book of Livy by the omission of difficult passages, being a First Latin Reading Book, with Grammatical Notes. Fcap. 8vo. 3s.

> This work is intended to supply the pupil with an easy Construing-book, which may at the same time be made the vehicle for instructing him in the rules of grammar and principles of composition. Here Livy tells his own pleasant stories in his own pleasant words. Let Livy be the master to teach a boy Latin, not some English collector of sentences, and he will not be found a dull one.

— A VOCABULARY AND EXERCISES on the "SEVEN KINGS OF ROME." Fcap. 8vo. 2s. 6d.

The Vocabulary and Exercises may also be had bound up with "The Seven Kings of Rome," price 5s.

MATHEMATICAL.

AIRY.—Works by **G. B. Airy**, Astronomer Royal :—

— ELEMENTARY TREATISE ON PARTIAL DIFFERENTIAL EQUATIONS. Designed for the use of Students in the University. With Diagrams. Crown 8vo. cloth, 5s. 6d.

> It is hoped that the methods of solution here explained, and the instances exhibited, will be found sufficient for application to nearly all the important problems of Physical Science, which require for their complete investigation the aid of partial differential equations.

AIRY.—Works by **G. B. Airy**—*Continued.*

— ON THE ALGEBRAICAL AND NUMERICAL THEORY of ERRORS of OBSERVATIONS, and the COMBINATION of OBSERVATIONS. Crown 8vo. cloth, 6s. 6d.

— UNDULATORY THEORY OF OPTICS. Designed for the use of Students in the University. New Edition. Crown 8vo. cloth, 6s. 6d.

— ON SOUND and ATMOSPHERIC VIBRATIONS. With the Mathematical Elements of Music. Designed for the use of Students of the University. Crown 8vo. 9s.

BAYMA.—THE ELEMENTS of MOLECULAR MECHANICS. By Joseph Bayma, S.J., Professor of Philosophy, Stonyhurst College. Demy 8vo. cloth, 10s. 6d.

BOOLE.—Works by **G. Boole, D.C.L., F.R.S.,** Professor of Mathematics in the Queen's University, Ireland :—

— A TREATISE ON DIFFERENTIAL EQUATIONS. New and Revised Edition. Edited by I. Todhunter. Crown 8vo. cloth, 14s.

> The author has endeavoured in this Treatise to convey as complete an account of the present state of knowledge on the subject of Differential Equations, as was consistent with the idea of a work intended primarily for elementary instruction. The earlier sections of each chapter contain that kind of matter which has usually been thought suitable to the beginner, while the later ones are devoted either to an account of recent discovery, or the discussion of such deeper questions of principle as are likely to present themselves to the reflective student in connexion with the methods and processes of his previous course.

— A TREATISE ON DIFFERENTIAL EQUATIONS. Supplementary Volume. Edited by I. Todhunter. Crown 8vo. cloth, 8s. 6d.

— THE CALCULUS OF FINITE DIFFERENCES. Crown 8vo. cloth, 10s. 6d.

> This work is in some measure designed as a sequel to the *Treatise on Differential Equations,* and is composed on the same plan.

BEASLEY.—AN ELEMENTARY TREATISE ON PLANE TRIGONOMETRY. With Examples. By R. D. BEASLEY, M.A., Head Master of Grantham Grammar School. Second Edition, revised and enlarged. Crown 8vo. cloth, 3s. 6d.

This Treatise is specially intended for use in Schools. The choice of matter has been chiefly guided by the requirements of the three days' Examination at Cambridge, with the exception of proportional parts in Logarithms, which have been omitted. About *Four Hundred* Examples have been added, mainly collected from the Examination Papers of the last ten years, and great pains have been taken to exclude from the body of the work any which might dishearten a beginner by their difficulty.

CAMBRIDGE SENATE-HOUSE PROBLEMS and RIDERS, WITH SOLUTIONS:—

1848—1851.—PROBLEMS. By FERRERS and JACKSON. 8vo. cloth. 15s. 6d.

1848—1851.—RIDERS. By JAMESON. 8vo. cloth. 7s. 6d.

1854.—PROBLEMS and RIDERS. By WALTON and MACKENZIE, 8vo. cloth. 10s. 6d.

1857.—PROBLEMS and RIDERS. By CAMPION and WALTON. 8vo. cloth. 8s. 6d.

1860.—PROBLEMS and RIDERS. By WATSON and ROUTH. Crown 8vo. cloth. 7s. 6d.

1864.—PROBLEMS and RIDERS. By WALTON and WILKINSON. 8vo. cloth. 10s. 6d.

CAMBRIDGE COURSE OF ELEMENTARY NATURAL PHILOSOPHY, for the Degree of B.A. Originally compiled by J. C. SNOWBALL, M.A., late Fellow of St. John's College. Fifth Edition, revised and enlarged, and adapted for the Middle-Class Examinations by THOMAS LUND, B.D., Late Fellow and Lecturer of St. John's College; Editor of Wood's Algebra, &c. Crown 8vo. cloth. 5s.

This work will be found suited to the wants, not only of University Students, but also of many others who require a short course of Mechanics and Hydrostatics, and especially of the Candidates at our Middle-Class Examinations.

CAMBRIDGE AND DUBLIN MATHEMATICAL JOURNAL. THE COMPLETE WORK, in Nine Vols. 8vo. cloth. £7 4s.
(Only a few copies remain on hand.)

CHEYNE.—AN ELEMENTARY TREATISE on the PLANETARY THEORY. With a Collection of Problems. By C. H. H. CHEYNE, B.A. Crown 8vo. cloth. 6s. 6d.

— THE EARTH'S MOTION of ROTATION. By C. H. H. CHEYNE, M.A. Crown 8vo. 3s. 6d.

CHILDE.—THE SINGULAR PROPERTIES of the ELLIPSOID and ASSOCIATED SURFACES of the Nth DEGREE. By the Rev. G. F. CHILDE, M.A., Author of "Ray Surfaces," "Related Caustics," &c. 8vo. 10s. 6d.

CHRISTIE—A COLLECTION OF ELEMENTARY TEST-QUESTIONS in PURE and MIXED MATHEMATICS; with Answers and Appendices on Synthetic Division, and on the Solution of Numerical Equations by Horner's Method. By JAMES R. CHRISTIE, F.R.S., late First Mathematical Master at the Royal Military Academy, Woolwich. Crown 8vo. cloth, 8s. 6d.

DALTON.—ARITHMETICAL EXAMPLES. Progressively arranged, with Exercises and Examination Papers. By the Rev. T. DALTON, M.A., Assistant Master of Eton College. 18mo. cloth. 2s. 6d.

DAY.—PROPERTIES OF CONIC SECTIONS PROVED GEOMETRICALLY. Part I., THE ELLIPSE, with Problems. By the Rev. H. G. DAY, M.A., Head Master of Sedbergh Grammar School. Crown 8vo. 3s. 6d.

DODGSON.—AN ELEMENTARY TREATISE ON DETERMINANTS, with their Application to Simultaneous Linear Equations and Algebraical Geometry. By C. L. DODGSON, M.A., Mathematical Lecturer of Christ Church, Oxford. Small 4to. cloth, 10s. 6d.

DREW.—GEOMETRICAL TREATISE on CONIC SECTIONS. By W. H. DREW, M.A., St. John's College, Cambridge. Third Edition. Crown 8vo. cloth, 4s. 6d.

> In this work the subject of Conic Sections has been placed before the student in such a form that, it is hoped, after mastering the elements of Euclid, he may find it an easy and interesting continuation of his geometrical studies. With a view also of rendering the work a complete Manual of what is required at the Universities, there have been either embodied into the text, or inserted among the examples, every book-work question, problem, and rider, which has been proposed in the Cambridge examinations up to the present time.

— SOLUTIONS TO THE PROBLEMS IN DREW'S CONIC SECTIONS. Crown 8vo. cloth, 4s. 6d.

FERRERS.—AN ELEMENTARY TREATISE on TRILINEAR CO-ORDINATES, the Method of Reciprocal Polars, and the Theory of Projections. By the Rev. N. M. FERRERS, M.A., Fellow and Tutor of Gonville and Caius College, Cambridge. Second Edition. Crown 8vo. 6s. 6d.

> The object of the author in writing on this subject has mainly been to place it on a basis altogether independent of the ordinary Cartesian system, instead of regarding it as only a special form of Abridged Notation. A short chapter on Determinants has been introduced.

FROST.—THE FIRST THREE SECTIONS of NEWTON'S PRINCIPIA. With Notes and Illustrations. Also a Collection of Problems, principally intended as Examples of Newton's Methods. By PERCIVAL FROST, M.A., late Fellow of St. John's College, Mathematical Lecturer of King's College, Cambridge. Second Edition. 8vo. cloth, 10s. 6d.

> The author's principal Intention is to explain difficulties which may be encountered by the student on first reading the Principia, and to illustrate the advantages of a careful study of the methods employed by Newton, by showing the extent to which they may be applied in the solution of problems; he has also endeavoured to give assistance to the student who is engaged in the study of the higher branches of Mathematics, by representing in a geometrical form several of the processes employed in the Differential and Integral Calculus, and in the analytical Investigations of Dynamics.

FROST and WOLSTENHOLME.—A TREATISE ON SOLID GEOMETRY. By PERCIVAL FROST, M.A., and the Rev. J. WOLSTENHOLME, M.A., Fellow and Assistant Tutor of Christ's College. 8vo. cloth, 18s.

> The authors have endeavoured to present before students as comprehensive a view of the subject as possible. Intending as they have done to make the subject accessible, at least in the earlier portion, to all classes of students, they have endeavoured to explain fully all the processes which are most useful in dealing with ordinary theorems and problems, thus directing the student to the selection of methods which are best adapted to the exigencies of each problem. In the more difficult portions of the subject, they have considered themselves to be addressing a higher class of students; there they have tried to lay a good foundation on which to build, if any reader should wish to pursue the science beyond the limits to which the work extends.

GODFRAY.—A TREATISE on ASTRONOMY, for the use of Colleges and Schools. By HUGH GODFRAY, M.A., Mathematical Lecturer at Pembroke College, Cambridge. 8vo. cloth. 12s. 6d.

> "We can recommend for its purpose a very good *Treatise on Astronomy* by Mr. Godfray. It is a working book, taking astronomy in its proper place in mathematical science. But it begins with the elementary definitions, and connects the mathematical formulæ very clearly with the visible aspect of the heavens and the instruments which are used for observing it."—*Guardian.*

— AN ELEMENTARY TREATISE on the LUNAR THEORY. With a brief Sketch of the Problem up to the time of Newton. By HUGH GODFRAY, M.A. Second Edition, revised. Crown 8vo. cloth. 5s. 6d.

HEMMING.—AN ELEMENTARY TREATISE on the DIFFERENTIAL AND INTEGRAL CALCULUS, for the use of Colleges and Schools. By G. W. HEMMING, M.A., Fellow of St. John's College, Cambridge. Second Edition, with Corrections and Additions. 8vo. cloth. 9s.

JONES and CHEYNE.—ALGEBRAICAL EXERCISES. Progressively arranged. By the Rev. C. A. JONES, M.A., and C. H. CHEYNE, M.A., Mathematical Masters of Westminster School. New Edition. 18mo. cloth, 2s. 6d.

> This little book is intended to meet a difficulty which is probably felt more or less by all engaged in teaching Algebra to beginners. It is that while new ideas are being acquired, old ones are forgotten. In the belief that constant practice is the only remedy for this, the present series of miscellaneous exercises has been prepared. Their peculiarity consists in this, that though miscellaneous they are yet progressive, and may be used by the pupil almost from the commencement of his studies. They are not intended to supersede the systematically arranged examples to be found in ordinary treatises on Algebra, but rather to supplement them.
>
> The book being intended chiefly for Schools and Junior Students, the higher parts of Algebra have not been included.

KITCHENER.—A GEOMETRICAL NOTE-BOOK, containing Easy Problems in Geometrical Drawing preparatory to the Study of Geometry. For the use of Schools. By F. E. KITCHENER, M.A., Mathematical Master at Rugby. 4to. 2s.

MORGAN.—A COLLECTION of PROBLEMS and EXAMPLES in Mathematics. With Answers. By H. A. MORGAN, M.A., Sadlerian and Mathematical Lecturer of Jesus College, Cambridge. Crown 8vo. cloth. 6s. 6d.

> This book contains a number of problems, chiefly elementary, in the Mathematical subjects usually read at Cambridge. They have been selected from the papers set during late years at Jesus college. Very few of them are to be met with in other collections, and by far the larger number are due to some of the most distinguished Mathematicians in the University.

PARKINSON.—Works by S. Parkinson, B.D., Fellow and Prælector of St. John's College, Cambridge :—

— AN ELEMENTARY TREATISE ON MECHANICS. For the use of the Junior Classes at the University and the Higher Classes in Schools. With a Collection of Examples. Third Edition, revised. Crown 8vo. cloth, 9s. 6d.

> The author has endeavoured to render the present volume suitable as a Manual for the junior classes in Universities and the higher classes in Schools. In the Third Edition several additional propositions have been incorporated in the work for the purpose of rendering it more complete, and the Collection of Examples and Problems has been largely increased.

— A TREATISE on OPTICS. Second Edition, revised. Crown 8vo. cloth, 10s. 6d.

> A collection of Examples and Problems has been appended to this work which are sufficiently numerous and varied in character to afford useful exercise for the student : for the greater part of them recourse has been had to the Examination Papers set in the University and the several Colleges during the last twenty years.

PHEAR.—ELEMENTARY HYDROSTATICS. With numerous Examples. By J. B. PHEAR, M.A., Fellow and late Assistant Tutor of Clare College, Cambridge. Fourth Edition. Crown 8vo. cloth, 5s. 6d.

"An excellent Introductory Book. The definitions are very clear; the descriptions and explanations are sufficiently full and intelligible; the investigations are simple and scientific. The examples greatly enhance its value."—*English Journal of Education.*

PRATT.—A TREATISE on ATTRACTIONS, LAPLACE'S FUNCTIONS, and the FIGURE of the EARTH. By JOHN H. PRATT, M.A., Archdeacon of Calcutta, Author of "The Mathematical Principles of Mechanical Philosophy." Third Edition. Crown 8vo. cloth, 6s. 6d.

PUCKLE.—AN ELEMENTARY TREATISE on CONIC SECTIONS and ALGEBRAIC GEOMETRY. With numerous Examples and hints for their Solution; especially designed for the use of Beginners. By G. H. PUCKLE, M.A., St. John's College, Cambridge, Head Master of Windermere College. Third Edition, enlarged and improved. Crown 8vo. cloth, 7s. 6d.

The work has been completely re-written, and a considerable amount of new matter has been added, to suit the requirements of the present time.

RAWLINSON.—ELEMENTARY STATICS. By G. RAWLINSON, M.A. Edited by EDWARD STURGES, M.A., of Emmanuel College, Cambridge, and late Professor of the Applied Sciences, Elphinstone College, Bombay. Crown 8vo. cloth, 4s. 6d.

Published under the authority of H. M. Secretary of State for use in the Government Schools and Colleges in India.

"This Manual may take its place among the most exhaustive, yet clear and simple, we have met with, upon the composition and resolution of forces, equilibrium, and the mechanical powers."—*Oriental Budget.*

REYNOLDS. — MODERN METHODS IN ELEMENTARY GEOMETRY. By E. M. REYNOLDS, M.A., Mathematical Master in Clifton College. Crown 8vo. 3s. 6d.

ROUTH.—AN ELEMENTARY TREATISE on the DYNAMICS of a SYSTEM of RIGID BODIES. With Examples. By EDWARD JOHN ROUTH, M.A., Fellow and Assistant Tutor of St. Peter's College, Cambridge; Examiner in the University of London. Crown 8vo. cloth, 10s. 6d.

SMITH.—A TREATISE on ELEMENTARY STATICS. By J. H. SMITH, M.A., Gonville and Caius College, Cambridge. 8vo. 5s. 6d.

SMITH.—Works by **Barnard Smith**, **M.A.**, Rector of Glaston, Rutlandshire, late Fellow and Senior Bursar of St. Peter's College, Cambridge:—

— ARITHMETIC and ALGEBRA, in their Principles and Application, with numerous Systematically arranged Examples, taken from the Cambridge Examination Papers, with especial reference to the Ordinary Examination for B.A. Degree. Tenth Edition. Crown 8vo. cloth, 10s. 6d.

This work is now extensively used in *Schools* and *Colleges* both *at home* and in the *Colonies*. It has also been found of great service for students preparing for the MIDDLE-CLASS AND CIVIL AND MILITARY SERVICE EXAMINATIONS, from the care that has been taken to elucidate the *principles* of all the Rules.

— ARITHMETIC FOR SCHOOLS. New Edition. Crown 8vo. cloth, 4s. 6d.

— COMPANION to ARITHMETIC for SCHOOLS. [*Preparing.*

— A KEY to the ARITHMETIC for SCHOOLS. Seventh Edition. Crown 8vo., cloth, 8s. 6d.

— EXERCISES in ARITHMETIC. With Answers. Crown 8vo. limp cloth, 2s. 6d. Or sold separately, as follows:—Part I. 1s.; Part II. 1s. ANSWERS, 6d.

These Exercises have been published in order to give the pupil examples in every rule of Arithmetic. The greater number have been carefully compiled from the latest University and School Examination Papers.

— SCHOOL CLASS-BOOK of ARITHMETIC. 18mo. cloth, 3s. Or sold separately, Parts I. and II. 10d. each; Part III. 1s.

— KEYS to SCHOOL CLASS-BOOK of ARITHMETIC. Complete in one Volume, 18mo., cloth, 6s. 6d.; or Parts I., II., and III. 2s. 6d. each.

— SHILLING BOOK of ARITHMETIC for NATIONAL and ELEMENTARY SCHOOLS. 18mo. cloth. Or separately, Part I. 2d.; Part II. 3d.; Part III. 7d. ANSWERS, 6d.

THE SAME, with Answers complete. 18mo. cloth, 1s. 6d.

— KEY to SHILLING BOOK of ARITHMETIC. 18mo. cloth, 4s. 6d.

— EXAMINATION PAPERS in ARITHMETIC. In Four Parts. 18mo. cloth, 1s. 6d. THE SAME, with Answers, 18mo. 1s. 9d.

— KEY to EXAMINATION PAPERS in ARITHMETIC. 18mo. cloth, 4s. 6d.

SNOWBALL.—PLANE and SPHERICAL TRIGONOMETRY. With the Construction and Use of Tables of Logarithms. By J. C. SNOWBALL. Tenth Edition. Crown 8vo. cloth, 7s. 6d.

TAIT and STEELE.—DYNAMICS of a PARTICLE. With Examples. By Professor TAIT and Mr. STEELE. New Edition. Crown 8vo. cloth, 10s. 6d.

> In this Treatise will be found all the ordinary propositions connected with the Dynamics of Particles which can be conveniently deduced without the use of D'Alembert's Principles. Throughout the book will be found a number of illustrative Examples introduced in the text, and for the most part completely worked out ; others, with occasional solutions or hints to assist the student, are appended to each Chapter.

TAYLOR.—GEOMETRICAL CONICS ; including Anharmonic Ratio and Projection, with numerous Examples. By C. TAYLOR, B.A., Scholar of St. John's College, Cambridge. Crown 8vo. cloth, 7s. 6d.

TEBAY.—ELEMENTARY MENSURATION for SCHOOLS. With numerous Examples. By SEPTIMUS TEBAY, B.A., Head Master of Queen Elizabeth's Grammar School, Rivington. Extra fcap. 8vo. 3s. 6d.

TODHUNTER.—Works by I. Todhunter, M.A., F.R.S., Fellow and Principal Mathematical Lecturer of St. John's College, Cambridge :—

— THE ELEMENTS of EUCLID for the use of COLLEGES and SCHOOLS. New Edition. 18mo. cloth, 3s. 6d.

— ALGEBRA for BEGINNERS. With numerous Examples. New Edition. 18mo. cloth, 2s. 6d.

— KEY to ALGEBRA for BEGINNERS. Crown 8vo., cl., 6s. 6d.

— TRIGONOMETRY for BEGINNERS. With numerous Examples. New Edition. 18mo. cloth, 2s. 6d.

> Intended to serve as an introduction to the larger treatise on *Plane Trigonometry*, published by the author. The same plan has been adopted as in the *Algebra for Beginners* : the subject is discussed in short chapters, and a collection of examples is attached to each chapter.

— MECHANICS for BEGINNERS. With numerous Examples. 18mo. cloth, 4s. 6d.

> Intended as a companion to the two preceding books. The work forms an elementary treatise on *Demonstrative Mechanics*. It may be true that this part of mixed mathematics has been sometimes made too abstract and speculative ; but it can hardly be doubted that a knowledge of the elements at least of the theory of the subject is extremely valuable even for those who are mainly concerned with practical results. The author has accordingly endeavoured to provide a suitable introduction to the study of applied as well as of theoretical Mechanics.

TODHUNTER.—Works by I. Todhunter, M.A.—*Continued.*

— A TREATISE on the DIFFERENTIAL CALCULUS. With Examples. Fourth Edition, Crown 8vo. cloth, 10s. 6d.

— A TREATISE on the INTEGRAL CALCULUS. Third Edition, revised and enlarged. With Examples. Crown 8vo. cloth, 10s. 6d.

— A TREATISE on ANALYTICAL STATICS. With Examples. Third Edition, revised and enlarged. Crown 8vo. cloth, 10s. 6d.

— PLANE CO-ORDINATE GEOMETRY, as applied to the Straight Line and the CONIC SECTIONS. With numerous Examples. Fourth Edition. Crown 8vo. cloth, 7s. 6d.

— ALGEBRA. For the use of Colleges and Schools. Fourth Edition. Crown 8vo. cloth, 7s. 6d.

This work contains all the propositions which are usually included in elementary treatises on Algebra, and a large number of *Examples for Exercise.* The author has sought to render the work easily intelligible to students without impairing the accuracy of the demonstrations, or contracting the limits of the subject. The Examples have been selected with a view to illustrate every part of the subject, and as the number of them is about *Sixteen hundred and fifty,* it is hoped they will supply ample exercise for the student. Each set of Examples has been carefully arranged, commencing with very simple exercises, and proceeding gradually to those which are less obvious.

— PLANE TRIGONOMETRY. For Schools and Colleges. Third Edition. Crown 8vo. cloth, 5s.

The design of this work has been to render the subject intelligible to beginners, and at the same time to afford the student the opportunity of obtaining all the information which he will require on this branch of Mathematics. Each chapter is followed by a set of Examples; those which are entitled *Miscellaneous Examples,* together with a few in some of the other sets, may be advantageously reserved by the student for exercise after he has made some progress in the subject. In the Second Edition the hints for the solution of the Examples have been considerably increased.

— A TREATISE ON SPHERICAL TRIGONOMETRY. Second Edition, enlarged. Crown 8vo. cloth, 4s. 6d.

This work is constructed on the same plan as the *Treatise on Plane Trigonometry,* to which it is intended as a sequel. Considerable labour has been expended on the text in order to render it comprehensive and accurate, and the Examples, which have been chiefly selected from University and College Papers, have all been carefully verified.

— EXAMPLES of ANALYTICAL GEOMETRY of THREE DIMENSIONS. Second Edition, revised. Crown 8vo. cloth, 4s.

— AN ELEMENTARY TREATISE on the THEORY of EQUATIONS. Second Edition, revised. Crown 8vo. cloth, 7s. 6d.

WILSON.—ELEMENTARY GEOMETRY. PART I. Angles, Triangles, Parallels, and Equivalent Figures, with the Application to Problems. By J. M. WILSON, M.A., Fellow of St. John's College, Cambridge, and Mathematical Master in Rugby School. Extra fcap. 8vo. 2s. 6d.

— A TREATISE on DYNAMICS. By W. P. WILSON, M.A., Fellow of St. John's College, Cambridge; and Professor of Mathematics in Queen's College, Belfast. 8vo. 9s. 6d.

WOLSTENHOLME.—A BOOK of MATHEMATICAL PROB-LEMS on subjects included in the Cambridge Course. By JOSEPH WOLSTENHOLME, Fellow of Christ's College, sometime Fellow of St. John's College, and lately Lecturer in Mathematics at Christ's College. Crown 8vo. cloth. 8s. 6d.

> CONTENTS: Geometry (Euclid).—Algebra.—Plane Trigonometry.—Conic Sections, Geometrical.—Conic Sections, Analytical.—Theory of Equations. —Differential Calculus.—Integral Calculus.—Solid Geometry—Statics.—Dynamics, Elementary.—Newton.—Dynamics of a Point.—Dynamics of a Rigid Body.—Hydrostatics.—Geometrical Optics.—Spherical Trigonometry and Plane Astronomy.

> In each subject the order of the Text-Books in general use in the University of Cambridge has been followed, and to some extent the questions have been arranged in order of difficulty. The collection will be found to be unusually copious in problems in the earlier subjects, by which it is designed to make the work useful to mathematical students, not only in the Universities, but in the higher classes of public schools.

SCIENCE.

AIRY.—POPULAR ASTRONOMY. With Illustrations. By G. B. AIRY, Astronomer Royal. Sixth and Cheaper Edition. 18mo. cloth, 4s. 6d.

> "Popular Astronomy in general has many manuals; but none of them super-sede the Six Lectures of the Astronomer Royal under that title. Its speciality is the direct way in which every step is referred to the observatory, and in which the methods and instruments by which every observation is made are fully described. This gives a sense of solidity and substance to astronomical statements which is obtainable in no other way."—*Guardian.*

GEIKIE.—ELEMENTARY LESSONS in PHYSICAL GEO-LOGY. By ARCHIBALD GEIKIE, F.R.S., Director of the Geological Survey of Scotland. [*Preparing.*

HUXLEY.—LESSONS in ELEMENTARY PHYSIOLOGY. With numerous Illustrations. By T. H. HUXLEY, F.R.S., Professor of Natural History in the Royal School of Mines. Second Edition. 18mo. cloth, 4s. 6d.

> "It is a very small book, but pure gold throughout. There is not a waste sentence, or a superfluous word, and yet it is all clear as daylight. It exacts close attention from the reader, but the attention will be repaid by a real acquisition of knowledge. And though the book is so small, it manages to touch on some of the very highest problems. .... The whole book shows how true it is that the most elementary instruction is best given by the highest masters in any science."—*Guardian.*

> "The very best descriptions and explanations of the principles of human physiology which have yet been written by an Englishman."—*Saturday Review.*

LOCKYER.—ELEMENTARY LESSONS in ASTRONOMY. With Coloured Diagram of the Spectra of the Sun, Stars, and Nebulæ, and numerous Illustrations. By J. NORMAN LOCKYER, F.R.A.S. 18mo. 5s. 6d.

OLIVER.—LESSONS IN ELEMENTARY BOTANY. With nearly Two Hundred Illustrations. By DANIEL OLIVER, F.R.S., F.L.S. Third Thousand. 18mo. cloth, 4s. 6d.

> "The manner is most fascinating, and if it does not succeed in making this division of science interesting to every one, we do not think anything can. .... Nearly 200 well executed woodcuts are scattered through the text, and a valuable and copious index completes a volume which we cannot praise too highly, and which we trust all our botanical readers, young and old, will possess themselves of."—*Popular Science Review.*

> "To this system we now wish to direct the attention of teachers, feeling satisfied that by some such course alone can any substantial knowledge of plants be conveyed with certainty to young men educated as the mass of our medical students have been. We know of no work so well suited to direct the botanical pupil's efforts as that of Professor Oliver's, who, with views so practical and with great knowledge too, can write so accurately and clearly."—*Natural History Review.*

ROSCOE.—LESSONS in ELEMENTARY CHEMISTRY, Inorganic and Organic. By HENRY ROSCOE, F.R.S., Professor of Chemistry in Owen's College, Manchester. With numerous Illustrations and Chromo-Litho. of the Solar Spectra. Ninth Thousand. 18mo. cloth, 4s. 6d.

> It has been the endeavour of the author to arrange the most important facts and principles of Modern Chemistry in a plain but concise and scientific form, suited to the present requirements of elementary instruction. For the purpose of facilitating the attainment of exactitude in the knowledge of the subject, a series of exercises and questions upon the lessons have been added. The metric system of weights and measures, and the centigrade thermometric scale, are used throughout the work.

> "A small, compact, carefully elaborated and well arranged manual."—*Spectator.*

MISCELLANEOUS.

ATLAS of EUROPE. GLOBE EDITION. Uniform in size with Macmillan's Globe Series, containing 48 Coloured Maps, on the same scale Plans of London and Paris, and a copious Index, strongly bound in half-morocco, with flexible back. 9s.

> NOTICE.—This Atlas includes all the Countries of Europe in a Series of Forty-eight Maps, drawn on the same scale, with an Alphabetical Index to the situation of more than 10,000 Places; and the relation of the various Maps and Countries to each other is defined in a general Key-Map.
>
> The Identity of scale in all the Maps facilitates the comparison of extent and distance, and conveys a just impression of the magnitude of different Countries. The size suffices to show the Provincial Divisions, the Railways and Main Roads, the Principal Rivers and Mountain Ranges. As a book it can be opened without the inconvenience which attends the use of a folding map.
>
> "In the series of works which Messrs. Macmillan and Co. are publishing under this general title (*Globe Series*) they have combined portableness with scholarly accuracy and typographical beauty, to a degree that is almost unprecedented. Happily they are not alone in employing the highest available scholarship in the preparation of the most elementary educational works; but their exquisite taste and large resources secure an artistic result which puts them almost beyond competition. This little atlas will be an invaluable boon for the school, the desk, or the traveller's portmanteau."—*British Quarterly Review.*

BATES and LOCKYER.—A CLASS BOOK of GEOGRAPHY, adapted to the recent Programme of the Royal Geographical Society. By H. W. BATES and J. N. LOCKYER, F.R.A.S.
> [*In the Press.*

CAMEOS from ENGLISH HISTORY. From Rollo to Edward II. By the Author of "The Heir of Redclyffe." Extra fcap. 8vo. 5s.

> "Contains a large amount of information in a concentrated form, and so skilfully and well is the adventurous, personal, and dramatic element brought out, that any boy of intelligence will find these narratives as fascinating as the most exciting fiction ever penned."—*London Review.*

EARLY EGYPTIAN HISTORY for the Young. With Descriptions of the Tombs and Monuments. New Edition, with Frontispiece. Fcap. 8vo. 5s.

> "Written with liveliness and perspicuity."—*Guardian.*
>
> "Artistic appreciation of the picturesque, lively humour, unusual aptitude for handling the childish intellect, a pleasant style, and sufficient learning, altogether free from pedantic parade, are among the good qualities of this volume, which we cordially recommend to the parents of inquiring and book-loving boys and girls."—*Athenæum.*
>
> "This is one of the most perfect books for the young that we have ever seen. We know something of Herodotus and Rawlinson, and the subject is certainly not new to us; yet we read on, not because it is our duty, but for very pleasure. The author has hit the best possible way of interesting any one, young or old."—*Literary Churchman.*

HOLE.—A GENEALOGICAL STEMMA of the KINGS of ENG-
LAND and FRANCE. By the Rev. C. HOLE. In One Sheet.
1s.

— A BRIEF BIOGRAPHICAL DICTIONARY. Compiled and
Arranged by CHARLES HOLE, M.A., Trinity College, Cambridge.
Second Edition, 18mo., neatly and strongly bound in cloth,
4s. 6d.

> The most comprehensive Biographical Dictionary in English,—containing
> more than 18,000 names of persons of all countries, with dates of birth and
> death, and what they were distinguished for.

> "An invaluable addition to our manuals of reference, and from its moderate
> price, it cannot fail to become as popular as it is useful."—*Times.*

> "Supplies a universal want among students of all kinds. It is a neat, com-
> pact, well printed little volume, which may go into the pocket, and should
> be on every student's table, at hand, for reference."—*Globe.*

HOUSEHOLD (A) BOOK OF ENGLISH POETRY. Selected
and arranged, with Notes, by R. C. TRENCH, D.D., Archbishop of
Dublin. Extra fcap. 8vo. 5s. 6d.

> "Remarkable for the number of fine poems it contains that are not found in
> other collections."—*Express.*

> "The selection is made with the most refined taste, and with excellent
> judgment."—*Birmingham Gazette.*

JEPHSON.—SHAKESPEARE'S TEMPEST. With Glossary and
Explanatory Notes. By the Rev. J. M. JEPHSON. 18mo. 1s. 6d.

> "His notes display a thorough familiarity with our older English literature,
> and his preface is so full of intelligent critical remark, that many readers
> will wish that it were longer."—*Guardian.*

OPPEN.—FRENCH READER. For the use of Colleges and
Schools. Containing a Graduated Selection from Modern Authors
in Prose and Verse; and copious Notes, chiefly Etymological.
By EDWARD A. OPPEN. Fcap. 8vo. cloth, 4s. 6d.

> "Mr. Oppen has produced a French Reader, which is at once moderate yet
> full, informing yet interesting, which in its selections balances the moderns
> fairly against the ancients. . . . The examples are chosen with taste and
> skill, and are so arranged as to form a most agreeable course of French
> reading. An etymological and biographical appendix constitutes a very
> valuable feature of the work."—*Birmingham Daily Post.*

A SHILLING BOOK of GOLDEN DEEDS. A Reading-Book for
Schools and General Readers. By the Author of "The Heir of
Redclyffe." 18mo. cloth.

> "To collect in a small handy volume some of the most conspicuous of these
> (examples) told in a graphic and spirited style, was a happy idea, and the
> result is a little book that we are sure will be in almost constant demand in
> the parochial libraries and schools for which it is avowedly intended."—
> *Educational Times.*

A SHILLING BOOK of WORDS from the POETS. By C. M. VAUGHAN. 18mo. cloth.

THRING.—Works by **Edward Thring, M.A.,** Head Master of Uppingham :—

— THE ELEMENTS of GRAMMAR taught in ENGLISH. With Questions. Fourth Edition. 18mo. 2s.

— THE CHILD'S GRAMMAR. Being the substance of "The Elements of Grammar taught in English," adapted for the use of Junior Classes. A New Edition. 18mo. 1s.

> The author's effort in these two books has been to point out the broad, beaten, every-day path, carefully avoiding digressions into the bye-ways and eccentricities of language. This work took its rise from questionings in National Schools, and the whole of the first part is merely the writing out in order the answers to questions which have been used already with success. Its success, not only in National Schools, from practical work in which it took its rise, but also in classical schools, is full of encouragement.

— SCHOOL SONGS. A collection of Songs for Schools. With the Music arranged for Four Voices. Edited by the Rev. E. THRING and H. RICCIUS. Folio. 7s. 6d.

DIVINITY.

EASTWOOD.—THE BIBLE WORD BOOK. A Glossary of Old English Bible Words. By J. EASTWOOD, M.A., of St. John's College, and W. ALDIS WRIGHT, M.A., Trinity College, Cambridge. 18mo. 5s. 6d.

HARDWICK.—A HISTORY of the CHRISTIAN CHURCH. MIDDLE AGE. From Gregory the Great to the Excommunication of Luther. By ARCHDEACON HARDWICK. Edited by FRANCIS PROCTER, M.A. With Four Maps constructed for this work by A. KEITH JOHNSTON. Second Edition. Crown 8vo. 10s. 6d.

— A HISTORY of the CHRISTIAN CHURCH during the REFORMATION. By ARCHDEACON HARDWICK. Revised by FRANCIS PROCTER, M.A. Second Edition. Crown 8vo. 10s. 6d.

MACLEAR.—Works by the **Rev. G. F. Maclear, B.D.**, Head Master of King's College School, and Preacher at the Temple Church :—

— A CLASS-BOOK of OLD TESTAMENT HISTORY. Fourth Edition, with Four Maps. 18mo. cloth, 4s. 6d.

> "A work which for fulness and accuracy of information may be confidently recommended to teachers as one of the best text-books of Scripture History which can be put into a pupil's hands."—*Educational Times.*

— A CLASS-BOOK of NEW TESTAMENT HISTORY : including the Connection of the Old and New Testament. With Four Maps. Second Edition. 18mo. cloth. 5s. 6d.

> "Mr. Maclear has produced in this handy little volume a singularly clear and orderly arrangement of the Sacred Story.... His work is solidly and completely done."—*Athenæum.*

— A SHILLING BOOK of OLD TESTAMENT HISTORY, for National and Elementary Schools. With Map. 18mo. cloth.

— A SHILLING BOOK of NEW TESTAMENT HISTORY, for National and Elementary Schools. With Map. 18mo. cloth.

— CLASS BOOK of the CATECHISM of the CHURCH of ENGLAND. Second Edition. 18mo. cloth, 2s. 6d. A Sixpenny Edition in the Press.

PROCTER.—A HISTORY of the BOOK of COMMON PRAYER : with a Rationale of its Offices. By FRANCIS PROCTER, M.A. Seventh Edition, revised and enlarged. Crown 8vo. 10s. 6d.

> In the course of the last twenty years the whole question of Liturgical knowledge has been reopened with great learning and accurate research, and it is mainly with the view of epitomizing their extensive publications, and correcting by their help the errors and misconceptions which had obtained currency, that the present volume has been put together.

— AN ELEMENTARY HISTORY of the BOOK of COMMON PRAYER. By FRANCIS PROCTER, M.A. Second Edition. 18mo. 2s. 6d.

> The author having been frequently urged to give a popular abridgment of his larger work in a form which should be suited for use in schools and for general readers, has attempted in this book to trace the History of the Prayer-Book, and to supply to the English reader the general results which in the larger work are accompanied by elaborate discussions and references to authorities indispensable to the student. It is hoped that this book may form a useful manual to assist people generally to a more intelligent use of the Forms of our Common Prayer.

PSALMS of DAVID Chronologically Arranged. By FOUR FRIENDS. An amended version, with Historical Introduction and Explanatory Notes. Crown 8vo., 10s. 6d.

> "It is a work of choice scholarship and rare delicacy of touch and feeling."—*British Quarterly.*

RAMSAY.—THE CATECHISER'S MANUAL; or, the Church Catechism illustrated and explained, for the use of Clergymen, Schoolmasters, and Teachers. By ARTHUR RAMSAY, M.A. Second Edition. 18mo. 1s. 6d.

SIMPSON.—AN EPITOME of the HISTORY of the CHRIST- IAN CHURCH. By WILLIAM SIMPSON, M.A. Fourth Edition. Fcap. 8vo. 3s. 6d.

SWAINSON.—A HAND-BOOK to BUTLER'S ANALOGY. By C. A. SWAINSON, D.D., Norrisian Professor of Divinity at Cambridge. Crown 8vo. 1s. 6d.

WESTCOTT.—A GENERAL SURVEY of the HISTORY of the CANON of the NEW TESTAMENT during the First Four Centuries. By BROOKE FOSS WESTCOTT, B.D., Assistant Master at Harrow. Second Edition, revised. Crown 8vo. 10s. 6d.

> The Author has endeavoured to connect the history of the New Testament Canon with the growth and consolidation of the Church, and to point out the relation existing between the amount of evidence for the authenticity of its component parts and the whole mass of Christian literature. Such a method of inquiry will convey both the truest notion of the connexion of the written Word with the living Body of Christ, and the surest conviction of its divine authority.

— INTRODUCTION to the STUDY of the FOUR GOSPELS. By BROOKE FOSS WESTCOTT, B.D. Third Edition. Crown 8vo. 10s. 6d.

> This book is intended to be an Introduction to the *Study* of the Gospels. In a subject which involves so vast a literature much must have been over- looked; but the author has made it a point at least to study the researches of the great writers, and consciously to neglect none.

— THE BIBLE in the CHURCH. A Popular Account of the Collection and Reception of the Holy Scriptures in the Christian Churches. Second Edition. By BROOKE FOSS WESTCOTT, B.D. 18mo. cloth, 4s. 6d.

> " Mr. Westcott has collected and set out in a popular form the principal facts concerning the history of the Canon of Scripture. The work is executed with Mr. Westcott's characteristic ability."—*Journal of Sacred Literature.*

WILSON.—AN ENGLISH HEBREW and CHALDEE LEXI- CON and CONCORDANCE to the more Correct Understanding of the English translation of the Old Testament, by reference to the Original Hebrew. By WILLIAM WILSON, D.D., Canon of Winchester, late Fellow of Queen's College, Oxford. Second Edition, carefully Revised. 4to. cloth, 25s.

> The aim of this work is, that it should be useful to Clergymen and all per- sons engaged in the study of the Bible, even when they do not possess a knowledge of Hebrew; while able Hebrew scholars have borne testimony to the help that they themselves have found in it.

BOOKS ON EDUCATION.

ARNOLD.—A FRENCH ETON ; or, Middle-Class Education and the State. By MATTHEW ARNOLD. Fcap. 8vo. cloth. 2*s.* 6*d.*

> "A very interesting dissertation on the system of secondary instruction in France, and on the advisability of copying the system in England."— *Saturday Review.*

— SCHOOLS and UNIVERSITIES on the CONTINENT. 8vo. 10*s.* 6*d.*

BLAKE.—A VISIT to some AMERICAN SCHOOLS and COLLEGES. By SOPHIA JEX BLAKE. Crown 8vo. cloth. 6*s.*

> "Miss Blake gives a living picture of the schools and colleges themselves, in which that education is carried on."—*Pall-Mall Gazette.*

> "Miss Blake has written an entertaining book upon an important subject ; and while we thank her for some valuable information, we venture to thank her also for the very agreeable manner in which she imparts it."—*Athenæum.*

> "We have not often met with a more interesting work on education than that before us."—*Educational Times.*

ESSAYS ON A LIBERAL EDUCATION. By CHARLES STUART PARKER, M.A., HENRY SIDGWICK, M.A., LORD HOUGHTON, JOHN SEELEY, M.A., REV. F. W. FARRAR, M.A., F.R.S., &c., E. E. BOWEN, M.A., F.R.A.S., J. W. HALES, M.A, J. M. WILSON, M.A., F.G.S., F.R.A.S., W. JOHNSON, M.A. Edited by the Rev. F. W. FARRAR, M.A., F.R.S., late Fellow of Trinity College, Cambridge ; Fellow of King's College, London ; Assistant Master at Harrow ; Author of "Chapters on Language," &c., &c. Second Edition. 8vo. cloth, 10*s.* 6*d.*

FARRAR.—ON SOME DEFECTS IN PUBLIC SCHOOL EDUCATION. A Lecture delivered at the Royal Institution. With Notes and Appendices. Crown 8vo. 1*s.*

THRING.—EDUCATION AND SCHOOL. By the Rev. EDWARD THRING, M.A., Head Master of Uppingham. Second Edition. Crown 8vo. cloth. 6*s.*

YOUMANS.—MODERN CULTURE: its True Aims and Requirements. A Series of Addresses and Arguments on the Claims of Scientific Education. Edited by EDWARD L. YOUMANS, M.D. Crown 8vo. 8*s.* 6*d.*

CAMBRIDGE :—PRINTED BY JONATHAN PALMER.